T0317680

Organic Nanochemistry

Organic Nanochemistry

From Fundamental Concepts to Experimental Practice

Yuming Zhao
Department of Chemistry
Memorial University
St. John's, NL, Canada

Contents

Preface

About ten years ago, I was asked by Mr. Jonathan T. Rose, senior editor of Wiley, to suggest some new topics for organic chemistry-related textbooks. A title of *Organic Nanochemistry* immediately came into my mind as I have been greatly appealed to and actively engaged in research of nanoscale organic materials throughout my entire academic career as an organic chemist. At Memorial University, I have been working as a faculty member in the Department of Chemistry since 2004, teaching various organic chemistry courses ranging from introductory organic chemistry, advanced organic synthesis, organic spectroscopy to physical organic chemistry, and supramolecular organic chemistry. My research and teaching experience motivated me to write a textbook to expound the intricate and fascinating aspects of nanotechnology in modern organic chemistry.

It took me a rather long period of time to think and decide on the structure and contents of such a textbook. "Nano" has been a buzzword popularly known in nearly all scientific and engineering disciplines nowadays. Within this framework, organic nanochemistry represents only a small portion where organic chemistry intersects with modern nanotechnology. Nonetheless, the breadth and complexity of organic nanochemistry are still substantial, owing to the rapid development and great achievements in relevant fields over the past few decades. To date, many new frontiers at the interface of organic chemistry and nanotechnology have been unveiled. They hold immense potential for future scientific discovery and technological innovations. With these said, it is impossible for a single textbook to cover all the exciting developments in modern organic nanochemistry. The choice of topics to write about under the title of *Organic Nanochemistry* is mainly based on my own perspectives, interests, and experiences.

I intend to have this textbook attract the readership primarily from senior undergraduate students and graduate students who study and carry out research in chemical and materials sciences. To many of them, the amazing world of organic nanochemistry has not yet been clearly demonstrated from other chemistry courses. This textbook will provide them with fundamentally important concepts, theories, and methodologies for organic nanochemistry as well as the captivating applications of organic nanochemistry in modern technology and daily life. I also expect this textbook to be a useful reference source for researchers and professionals who seek to expand their knowledge and grasp the potential of organic nanochemistry.

There are totally six chapters in this textbook. The first chapter focuses on key concepts and theories of organic chemistry, which are essential to understand the fundamental properties of organic molecular and supramolecular systems. In Chapter 2, a wide range of useful synthetic methodologies for the synthesis and functionalization of organic nanomaterials are described and illustrated with examples and mechanisms. Chapter 3 is focused on the supramolecular aspects in organic nanochemistry, especially the well-developed disciplines of host–guest chemistry and organic self-assembly chemistry. Chapter 4 deals with a unique class of carbon-based nanomaterials, namely carbon nanoallotropes. Herein, the chemistry and application of exotic carbon nanomaterials emerged in recent decades, including fullerenes, carbon nanotubes, graphenes, and molecular nanocarbons, are systematically discussed. Chapter 5 provides the fundamental theories and case studies for the construction and testing of molecular devices and molecular machines. The examples discussed in this chapter showcase the power of organic chemistry in tuning and engineering of ultraminiaturized devices and machinery at the molecular level. Finally, state-of-the-art computational modeling methods for understanding and prediction of the properties of nanoscale organic systems, ranging from small molecules to large supramolecular materials, are introduced in Chapter 6. Various theories suitable for nanoscale simulations, including quantum mechanics, semiempirical quantum mechanics, and molecular dynamics theories, are discussed at an introductory level. Computational examples are provided, allowing interested readers to grasp essential modeling techniques for organic nanochemistry.

Overall, I anticipate this book will take the reader on a journey from familiar chemical concepts and principles to cutting-edge research of nanoscience and technology. The scope of topics and examples included in this book are only "the tip of the iceberg" that highlights the phenomenal achievements in organic nanochemistry over the past few decades. It is my sincere hope that this book will inspire and encourage readers, especially young students and researchers, to delve into the fascinating world of organic nanochemistry. As we unlock the secrets at the bottom of this world, we uncover the remarkable properties and possibilities that will arise in the future. I therefore invite you to embark on this captivating voyage and discover the transformative power of organic nanochemistry.

I would like to say that writing a book like this is an exercise in solidary endurance, which requires long periods of time alone to research, think, and put thoughts into words. The job was particularly daunting and arduous at the beginning stage. I am extremely grateful to my colleague and dear friend, Prof. Christopher Flinn, who spent his valuable time in reading through the manuscript and discussing with me frequently. Talking with Chris was always helpful, encouraging, and delightful. I am also very grateful to all the graduate and undergraduate students I have taught and supervised at Memorial University. Their curiosity and questions about chemistry and nanoscience greatly helped me formulate the ideas of many topics and case studies in this book. Among them, I am particularly thankful to a very talented graduate student, Ms. Parinaz Salari, who lent her creative skills to design and illustrate the cover image of this book. I want to express my deep appreciation to the support by the Wiley publishing team during the writing of this book. Last but not the least, I sincerely thank my wife, Lidan Tao. Words cannot fully capture the depth of my gratitude for the love, care, and unwavering support that Lidan brings into my life.

Yuming Zhao
St. John's, Newfoundland
June 2023

1

Fundamental Concepts

1.1 Introduction to Nanoscience and Nanotechnology

The word "nano" originated from the ancient Greek νᾶνος (nânos), meaning dwarf. In modern science and technology, nano is specifically used as an SI prefix, referring to the multiplying factor of 10^{-9}. For example, an anti-HIV/AIDs drug can be said to give an inhibitory concentration of a few nanomoles per liter (10^{-9} mol L^{-1}), an organic fluorescent molecule shows a fluorescence lifetime of tens of nanoseconds (10^{-9} second), and the third carbon allotrope, Buckminsterfullerene C_{60} (bucky ball), has a diameter of 1 nanometer (10^{-9} meter) as illustrated in Figure 1.1. Today the use of "nano" has gone far beyond its numerical meaning. In the academic world, nano has become a buzzword spanning the fields of chemistry, physics, biology, medicine, and engineering, where it is used to describe studies of advanced molecular, macromolecular, and supramolecular materials with the dimension of nanometers. In the industrial sector, many nanotechnology companies have been established to deal with the development and commercialization of nanometer-sized materials and devices for special purposes and tasks. The National Nanotechnology Initiative (NNI) in the US has defined nanotechnology as a science, technology, and engineering conducted at the scale of 1–100 nm. Unique properties and functions would arise from materials on such a small scale, which are often unprecedented and considerably more advantageous in comparison to conventional bulk materials.

So, how did nanotechnology start? The production and manipulation of nanoscale materials in fact have a surprisingly long history. Take the famous ancient artifact, the *Lycurgus cup*, as an example. This cup is an impressive Roman treasure made in about AD 400, which was named for its depiction of King Lycurgus of Thrace entangled in grape vines. The *Lycurgus cup* is well known because of its color-changing glass. When light is shone upon it, the cup displays

Organic Nanochemistry: From Fundamental Concepts to Experimental Practice, First Edition. Yuming Zhao.

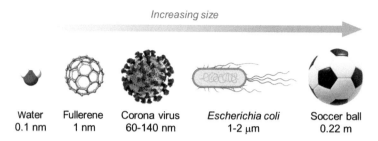

Figure 1.1 Illustration of various objects at the meter, micrometer, and nanometer scales. Credit: Alissa Eckert, MS; Dan Higgins, MAM / Wikimedia Commons / Public Domain.

a jade green color. When light is shone through it, however, the cup turns into a brilliant red color. The reason for this color-changing property is due to the gold–silver alloyed nanoparticles that are distributed in the glass. These nanoparticles scatter reflected light and back-illuminated light in different ways. Another example of ancient nanotechnology is the *Damascus steel swords* from the Middle East, which were made between AD 300 and AD 1700. They are known for their superior strength, shatter resistance, and exceptionally sharp cutting edge, owing to the use of the so-called wootz steel in their blade making. A recent scientific discovery disclosed that the wootz steel is full of carbon nanotubes, which are a class of appealing nanomaterials discovered by scientists in 1991. Nowadays, those artifacts would be included in a sub-branch of nanotechnology, known as nanocomposites. Beyond a doubt, those ancient artisans possessed masterful skills and empirical knowledge which enabled them to fabricate such stunning artifacts, but the presence and roles of the nanomaterials were not consciously known by them, neither were their related chemical and physical principles.

Many consider that the year of 1959 marked the inception of the concept of modern nanotechnology. In the December of that year, a visionary American physicist, Richard Feynman, who was the Nobel Prize laureate in Physics in 1965, gave a lecture at an annual meeting of the American Physical Society at Caltech, entitled *There's Plenty of Room at the Bottom* [1]. In this famous lecture, Feynman envisioned a day when devices and machines could be miniaturized in such a way that huge amounts of information could be stored in extremely small spaces, while machines could be fabricated and compacted together at a much smaller scale. Feynman's futuristic ideas sparked the beginning of the modern nanotechnology, although Feynman himself never used the term "nanotechnology" in his lectures. It was Prof. Norio Taniguchi of the Tokyo University of Science who coined the term "**nano-technology**" fifteen years later. In 1974, Taniguchi published a paper entitled *On the Basic Concept of Nanotechnology*, in which he wrote "Nanotechnology mainly consists of the processes of separation, consolidation, and deformation of materials by one atom or one molecule" [2]. This term was then adopted and greatly promoted by an American engineer, Eric Drexler, in his popular book *Engines of Creation: The Coming Era of Nanotechnology* published in 1986 [3]. In this book, Drexler imagined numerous future technologies, including an unprecedented class of tiny machines, which he termed **molecular assemblers**. He predicted that these assemblers would have the ability to precisely build objects in an atom-by-atom manner.

The era from 1980s to 1990s was full of exciting experimental discoveries and achievements in nanotechnology. In 1981, Gerd Binnig and Heinrich Rohrer at IBM Zurich Research Laboratories in Switzerland developed the first working scanning tunneling microscope (STM), for which they won the Nobel Prize in Physics in 1986. In 1985, these IBM researchers developed the technique called atomic force microscopy (AFM), which added another powerful tool for imagining and manipulating nanoscale objects on various substrates. A *tour de force* in precise positioning of atoms was achieved by IBM researchers Don Eigler and Erhard K. Schweizer in 1989, who used STM to create a tiny IMB logo with only 35 xenon atoms (see Figure 1.2). A range of nanoscale materials were discovered and developed during this period, which opened many sub-branches in modern nanotechnology.

Nanotechnology is not an individual discipline in science and engineering. Unlike the traditional disciplines (e.g., mathematics, chemistry, physics, and biology), nanotechnology is more like a technological hub that gathers a vast array of research under its umbrella. Research involving nanoscale materials has permeated almost every classical division of science and technology, making it highly multidisciplinary and without a clear boundary. Therefore, it is very hard to define the skills and precise type of backgrounds required to be a "nanotechnologist." Nanotechnology broadly covers engineering, chemistry, biology, medicine, computer science, theoretical simulations, devices and structures fabrication, just to name a few. In most nano-related studies, the preparation of nanoscale materials and precise control over their dimensions and shapes take center stage. To achieve these goals, special methods for nanofabrication are needed. In general, nanofabrication can be carried out using two different approaches, namely **top-down** and **bottom-up**. The top-down approach starts with the manipulation

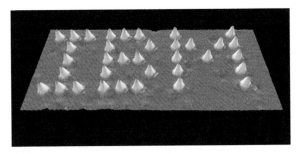

Figure 1.2 An IBM logo formed by positioning 35 xenon atoms on a nickel (110) surface. Credit IBM.

of bulk materials and structures. For example, the fabrication of a computer chip is done by "engraving" the surface of a piece of single-crystal silicon using a technique known as photolithography as the key step. Through the top-down approach, miniaturized devices are "chiseled out" of the bulk material in a precisely controlled manner, with the original integrity of the bulk material (e.g., crystallinity and long-range order) still retained. The bottom-up approach produces nanomaterials and devices through the self-assembly or chemical synthesis of certain nanoscale building blocks. The self-assembling process is dictated by specific chemical and/or physical forces (e.g., metal ligand coordination, hydrogen bonding, and π–π stacking) to form defined nanostructures. The building blocks can be obtained from naturally existing materials (e.g., DNA, lipids, carbohydrates) or prepared by chemical synthesis (e.g., synthetic molecules and polymers). The biological world is full of events utilizing the bottom-up approach, such as protein synthesis and cellular growth. Inspired by these natural materials and events, enormous efforts have been dedicated to applying various molecular functions and supramolecular forces to create nanostructures and materials.

1.2 Nanochemistry in Action

As prophesized by Richard Feynman, *there's plenty of room at the bottom*. Atoms and molecules are at the bottom of our physical world and can be manipulated and controlled. The knowledge and techniques for dealing with the assembly of atoms and molecules have long existed in the field of chemistry. According to the definition provided by Britannica, chemistry is "the science that deals with the properties, composition, and structure of substances (defined as elements and compounds), the transformations they undergo, and the energy that is released or absorbed during these processes". Jean-Marie Lehn, a 1987 Nobel Prize laureate in Chemistry, stated that "the science of chemistry is not just about discovery. It is also, and especially, about creation. It is an art of the complexification of matter. To understand the logic of the latest discoveries in nanochemistry, we have to take a 4-billion year leap back in time". So, it is obvious that chemists have been equipped with the required tools (knowledge of the properties of molecules including their reactions) to design and prepare nanoscale materials from the bottom up.

The intermarriage of chemistry and nanoscience has given rise to the vibrant interdisciplinary area of **nanochemistry**. One of the main focuses of nanochemistry is placed on the design and synthesis of molecular building blocks for the construction of various nanostructures. In this regard, the basic concepts and principles of nanochemistry do not differ too much from those encountered in traditional covalent and non-covalent chemistry. For example, the synthesis of a functional unit in a molecular machine can be achieved through the same bond-forming reactions that are useful in the synthesis of natural products or pharmaceutical compounds. The assembly of a nanoscale supramolecular structure may utilize the hydrogen bonding interactions in a manner similar to the Watson-Crick base pairing that typically exists in DNA and RNA molecules. What have proven to be more important for nanochemistry are the following: the creation of novel nanomaterials useful for technological applications, the development of new methods for more efficient materials synthesis and better characterization, and the understanding of the performances of nanomaterials at the molecular and supramolecular levels. In this respect, the existing chemical and biological literature is of great value for nano researchers to learn from and to be inspired.

Over the past few decades, nanochemistry has yielded fruitful results that have made profound impacts in our everyday lives and social development. Many advanced nanoscale organic and inorganic materials have found applications in the fields of advanced electronics and optoelectronics. New nanoelectronic materials are being used to replace the traditional conducting and semiconducting materials for improved device performances. For example, carbon nanotubes can be used in the fabrication of flexible displays and stretchable electronic devices through printing methods. Graphene serves as a robust and transparent electrode material in the production of new-generation touch screens. In the food industry, nanostructured materials have already been used for food packaging, functional food development, food safety, and shelf-life extension of food products. In solar energy technology, the use of nanoparticles such as quantum dots has helped to achieve enhanced absorption and sunlight energy capture, resulting in the development of solar cells that perform more efficiently. Nanoparticles such as titanium oxide (TiO_2) can be coated on the surface of glass panels or other solid substrates to form self-cleaning nanocoatings. Photocatalytic reactions enable such nanoparticles to clean dirt and grime on the surface. In a similar way, functional nanoparticles have been designed and engineered to act as efficient photocatalysts for water splitting, producing hydrogen as a clean energy source. Nanotechnology has also made significant impacts in environmental control and remediation. For instance, toxic organic pollutants in wastewater can be conveniently degraded under sunlight with the aid of photocatalytic nanoparticles. Nanoadsorbents can effectively remove toxic heavy metals from wastewater. In the field of medicine, nanomaterials have been widely used in diagnosis, controlled drug delivery, and regenerative medicines. For example, polymer-based nanocontainers are designed to show responses to external stimuli (temperature changes, light, ultrasound), which trigger the release of diagnostic molecules or drugs carried inside when they reach the targeted tissues or cells. Recently, "nanobots" have been fabricated to serve as miniature surgeons to repair damaged cells and modify intracellular structures in a highly precise manner. Such technology is expected to revolutionize the therapeutic methods for chronic diseases such as cancer. Even our daily lives have felt the benefits of nanochemistry. For example, smart nanocoated fabrics have been developed to exhibit amazing resistance to water, stain, and wrinkling. Nanocomposite materials have allowed sports equipment to be stronger, lighter, and more durable.

1.3 Structures and Covalent Bonding of Organic Compounds

1.3.1 Localized Covalent Bonds and Lewis Structures

In the molecular world, the distribution of electrons in molecules plays a vital role in controlling the structural, physical, and chemical properties of molecule-based materials. So, it is essential to establish a deep level of understanding of this quintessential property prior to carrying out various kinds of chemical research. **Lewis structures** are a simple and powerful approach to describe the structures of molecules, particularly organic molecules. The concept was first formulated by G. N. Lewis in 1916, who proposed that chemical bonds resulted from the sharing of electron pairs between atoms [4]. For the second-row atoms, Lewis stipulated that the most stable structures possess eight valence shell electrons, known as the **rule of eight** or **octet rule**. Following this rule, Lewis structures containing second-row atoms with more than eight valence electrons are considered too unstable to exist, and therefore should be avoided in the structural drawing.

Figure 1.3 illustrates the Lewis structures for various molecules containing second-row atoms and hydrogens. It can be seen that covalent bonds are drawn as lines, each of which represents a pair of shared valence electrons. Non-bonding valence electrons are drawn as either a pair of dots to represent an electron lone pair or a single dot for a free radical center. Positive and negative formal charges need to be assigned to cationic, anionic, and charge-separated species. The rules for drawing Lewis structures are not derived from rigorous quantum theories; however, they serve as an intuitive and powerful tool to account for electrons and charges in molecules. Straightforward pictures of electron distribution in molecules can be perceived from the Lewis structures.

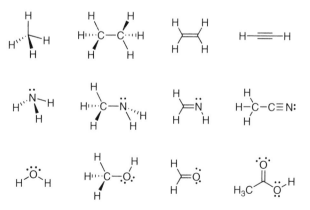

Figure 1.3 Lewis structures of simple molecules containing hydrogen, carbon, nitrogen, and oxygen atoms.

One major limitation of the Lewis structures is that they cannot provide information about the geometry of the molecules they depict (e.g., the molecular shapes and spatial arrangements of their constituent atoms). Nevertheless, this deficiency can be compensated by applying the **valence shell electron pair repulsion** (VSEPR) theory [5–7]. The VSEPR (pronounced "vɛspər") theory is based on the consideration of the Pauli Exclusion Principle and electrostatic repulsion of valence electrons (Coulomb's Law). Simply put, valence electron pairs in a molecule repel one another, and the stable structure of a molecule is a result of the minimization of these repulsions. The repulsion model can be readily applied to the AX_n type of molecular systems to predict the geometries as outlined in Figure 1.4. Herein, A is a central atom that is covalently bonded to several ligands (X_n). Assuming that there are no nonbonding electron pairs present, so only the A–X bonds contribute to the electrostatic repulsion within the system. If we treat each of the A–X bond as a point charge, the optimal arrangement of an AX_n system can be predicted using a model of charges on a sphere. As shown in Figure 1.4, to obtain the minimum mutual charge repulsion, the geometric outcomes for AX_n systems should be linear (n = 2), trigonal planar (n = 3), tetrahedral (n = 4), trigonal bipyramidal (n = 5), and octahedral (n = 6).

With the aid of the VSEPR theory, the geometry of simple organic molecules can be readily predicted. As shown in Figure 1.5, the molecule of carbon dioxide (CO_2) is predicted to take a linear shape, considering the electrostatic repulsion occurring between the two sets of C=O bonds. In a similar way, the geometries of borane (BH_3) and methane (CH_4) are predicted to take a trigonal planar and a tetrahedral shape, respectively. As a result, the bond angles in these two molecules are 120° and 109.5°, respectively. If we take steric effects into consideration, the experimentally observed bond angles which are slightly deviated from the ideal VSEPR predicted geometries can be reasonably explained. For example, the

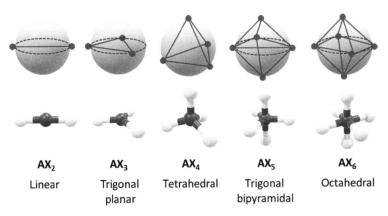

AX_2	AX_3	AX_4	AX_5	AX_6
Linear	Trigonal planar	Tetrahedral	Trigonal bipyramidal	Octahedral

Figure 1.4 Predicted geometries for various AX_n systems according to the VSEPR theory.

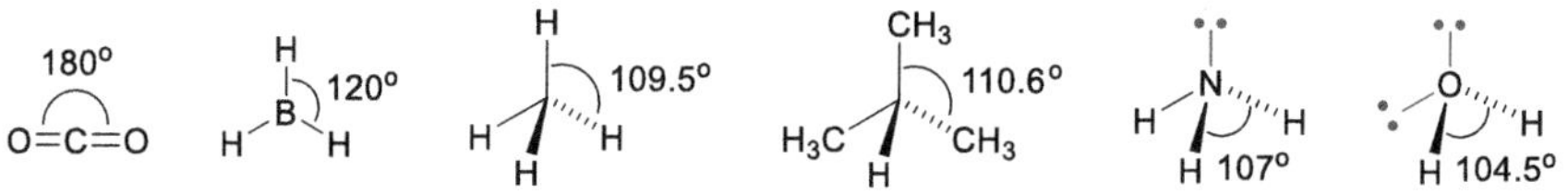

Figure 1.5 Bond angles in various simple molecules.

C–C–C bond angle in isobutane is determined to be 110.6°, which is slightly larger than the angle of 109.5° for an ideal tetrahedral arrangement. This outcome can be explained by that the repulsion between two C–C bonds is greater than that between a C–H bond and a C–C bond, since the methyl group is bulkier than hydrogen. If non-bonding electron lone pairs are present in the valence shell of a molecule, they should be counted as a repulsive domain similar to a covalent bond. Note that the lone pair electrons are distributed closer to the nucleus than the bonding electron pairs. As such, they tend to give stronger repulsion to adjacent electron pairs than the bond electron pair does. The repulsive forces in a molecule thus show a decreasing trend as follows: lone pair–lone pair > lone pair–bond pair > bond pair–bond pair. These notions can be applied to better rationalize the molecular geometries of ammonia (NH_3) and water (H_2O), where the H–N–H and H–O–H angles are 107° and 104.5°, respectively.

Besides using Lewis structural formulas and VSEPR theory, molecular structures can be more accurately predicted through quantum mechanics (QM) calculations. One classical approach that was rooted in the notion of electron pair bonding is called the **valence bond** (VB) theory [8]. In 1927 Heitler and London applied this method to calculate the molecule of dihydrogen. Their results indicated that the two hydrogen nuclei show a certain distance at an energy minimum, with electron density accumulated in between them. In fact, the VB theory gave rise to the notions of covalent bonds that are equivalent to the theory of electron pair sharing proposed by G. N. Lewis. Further application of the VB theory to simple molecules, particularly those containing second-row elements, led to the concept of **hybridization**. Linus Pauling developed this concept by mathematically mixing atomic orbitals together and then dividing the sum into equivalent parts [9]. Each part is called a **hybrid orbital**. Using the hybridization model, the structures and orbital properties of molecules can be conveniently interpreted. For example, in methane molecule (Figure 1.6) the central carbon atom has its four atomic orbitals ($2s$, $2p_x$, $2p_y$, $2p_z$) participating in hybridization to give four equivalent hybrid orbitals, each of which contains 25% s character and 75% p character. These hybrid orbitals are hence referred to as the sp^3 hybrid orbitals. The carbon atom of methane was thus proposed to use four equivalent sp^3 hybrid orbitals to overlap with the 1s orbitals of four surrounding hydrogen atoms, respectively, forming four equivalent C–H bonds in an ideal tetrahedral arrangement. In the same way, the sp^3 hybridized model can be applied to molecules carrying lone pair electrons (e.g., ammonia and water, Figure 1.6), and the outcomes are consistent with the predictions of the VSEPR theory.

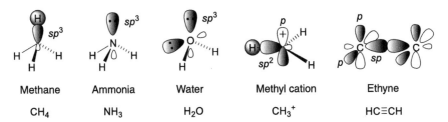

Figure 1.6 Hybrid orbitals in various molecular species.

Molecular species involving unsaturated structures can be rationalized by other hybridization states. For instance, the structure of methyl cation (CH_3^+) contains three equivalent C–H bonds formed through the overlaps of three sp^2 hybrid orbitals of carbon and three 1s orbitals of hydrogen. The $2p_z$ orbital of methyl cation does not participate in the hybridization and remains empty. This orbital picture well accounts for the electron deficiency of the methyl cation. In the molecule of ethyne (HC≡CH), the carbon–carbon triple bond constitutes a σ bond (overlap of two sp hybrid orbitals of carbon) and two sets of π bonds perpendicularly oriented toward each other. Each of the π bonds is formed by overlapping two unhybridized 2p orbitals of carbon in a shoulder-by-shoulder manner. Like the Lewis structures and VSEPR theory, the hybridization-derived σ/π model also offers a convenient tool to rationalize the geometry and bonding properties of a wide range of organic compounds.

It is worth noting that hybridization is only a theoretical model to account for certain experimentally observed outcomes (e.g., bond lengths, bond angles). By no means it should be treated as a fact or experimental observable. Take methane as an example, the hybridization model predicts that the four valence electrons of carbon are located in four equivalent C–H bonds, therefore these electrons should be of the same energy. Photoelectron spectroscopic analysis indicates that the four valence electrons of carbon in methane show different binding energies (12.7 eV and 23.0 eV), which contradict the hybridization model. Clearly, the hybridization model is flawed and fails to explain the electronic energies in this case. Nevertheless, the concept of hybridization is very useful in understanding certain aspects of molecular properties, and hence it is still widely used by chemists today.

Since the hybridization theory is based on a purely mathematic treatment of atomic orbitals, the hybridization states can be various in addition to the commonly seen sp^3, sp^2, and sp types. Hybridization state can be correlated to the interorbital angle (θ) as depicted in Figure 1.7A. Herein, a hybridization index (λ) can be derived from a known interorbital angle, and the hybridization state of an orbital is expressed by sp^n, where n = λ^2. Following this method, the number n can be either an integer or a real number, depending on the value of θ. This treatment leads to the so-called **variable hybridization theory**, which can be more flexibly applied to strained molecular systems where internuclear bond

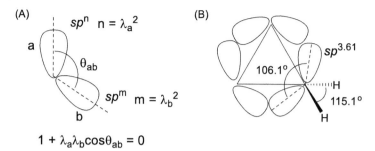

(A)

sp^n $n = \lambda_a^2$

a

θ_{ab}

sp^m $m = \lambda_b^2$

b

$1 + \lambda_a\lambda_b cos\theta_{ab} = 0$

(B)

$sp^{3.61}$

106.1°

H

115.1°

H

Figure 1.7 (A) Calculation of hybridization index (λ). (B) Description of the bonding properties of cyclopropane using the variable hybridization model.

angles significantly deviate from the ideal values. For example, the molecule of cyclopropane adopts a triangular shape, in which the C–C–C bond angle is 60°. This bond angle reflects the angle between the two lines crossing the positions of carbon nuclei, referred to as the **internuclear bond angle**. The angle between the two orbital lobes as depicted in Figure 1.7A is called **interorbital bond angle** (θ_{ab}), which can be obtained from the equation, $1 + \lambda_a\lambda_b cos\theta = 0$, where λ_a and λ_b are the hybridization indexes of orbital lobes a and b. In the molecule of cyclopropane, the H–C–H bond angle was experimentally determined as 115.1°. According to the variable hybridization theory, each carbon atom of cyclopropane uses two $sp^{2.39}$ hybrid orbitals to form two equivalent C–H bonds. This will leave the carbon atom two $sp^{3.61}$ hybrid orbitals for the formation of two C–C bonds. The interorbital angle of the C–C–C bond can therefore be calculated as 106.1° using the above equation. This interorbital angle is much greater than the internuclear C–C–C bond angle, pointing to a curved shape of the C–C bond (also known as the banana bond or tau bond). The curved or bent C–C bonds of cyclopropane thus result in a large strain energy of 27.6 kcal mol^{-1}. Also, these C–C bonds show a high degree of π-bond (alkene-like) character.

In addition to molecular geometry, the hybridization state can also be used to interpret other chemical properties. For example, the relative acidity of the C–H groups in alkanes, alkenes, and terminal alkynes shows a decreasing order with increasing degree of unsaturation. A saturated alkane molecule has a pK_a value more than 60. For an alkene molecule, the pK_a value is decreased to about 43, while for a terminal alkyne the pK_a is considerably lowered to 25. The dramatically changed C–H acidity can be correlated to the hybridization states of different carbons. In a second-row atom the 2s orbital stabilizes the valence electrons much better than the 2p orbital does. Increasing the s character in a hybrid orbital should result in stronger electron attraction (i.e., enhanced electronegativity). For this reason, the degree of polarization of a C–H bond increases with increasing s character in the hybrid orbital that forms the bond. As such, the C–H acidity follows a decreasing trend of alkyne (sp) > alkene (sp^2) > alkane (sp^3).

1.3.2 Delocalized Covalent Bonds, Conjugation, and Resonance Theory

Early studies of the dihydrogen molecule using the VB theory showed that the two bonding electrons are allowed to exchange their positions with respect to the two H nuclei, which can be expressed by two wavefunctions (Φ_a and Φ_b). Heisenberg proposed that electrons are indistinguishable particles. So, the two wavefunctions can be linearly combined in two different ways ($\Phi_a + \Phi_b$) and ($\Phi_a - \Phi_b$), leading to a splitting of energy between them. Heisenberg called this term **resonance**, which was coined based on a classical analogy of two oscillators that resonate at the same frequency.

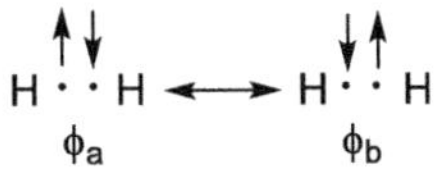

$$H \cdot\cdot H \longleftrightarrow H \cdot\cdot H$$
$$\phi_a \qquad\qquad \phi_b$$

The concept of resonance [10] is useful for interpreting the electron distribution of π-conjugated molecular systems, where bonding electrons are delocalized. For example, the molecule of benzene can be described by two equivalent Lewis structures of cyclohexatriene (Kekulé structure) shown in Figure 1.8. Note that a double-headed arrow is used in resonance scheme drawing to indicate a relationship of resonance. Each of the Lewis structures is called a **resonance contributor** and it reflects partial properties of the molecule. The overall structure of a molecule can be treated as a *hybrid* of its resonance contributors.

The resonance scheme offers an intuitive tool to explain the bonding properties and charge distributions in a delocalized system. Taking the allyl cation as an example (Figure 1.8), the positive charge is shown at the two terminal allyl carbons, respectively, in the resonance scheme. Since the two resonance contributors of allyl cation are equivalent, the two terminal carbon atoms are predicted to hold an equal amount of positive charge and the two C–C bonds show an equal bond distance and bond order. In a similar way, the electronic and bonding properties of an allyl anion can be rationalized.

It is important to know that the Lewis structures drawn in a resonance scheme represent various characteristics of the same molecular species, and they should not be confused with separate chemical species in equilibrium. From the quantum

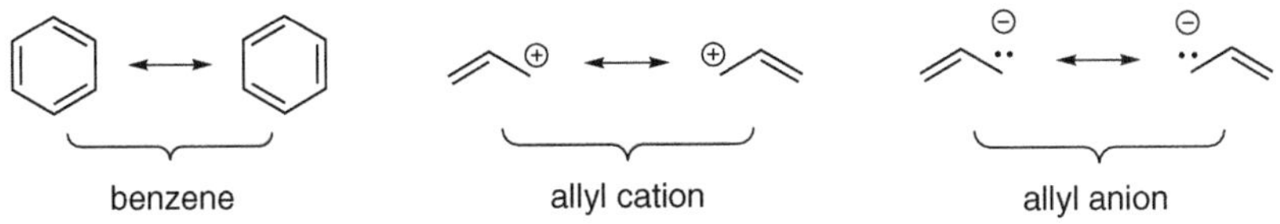

| benzene | allyl cation | allyl anion |

Figure 1.8 Resonance schemes of π-conjugated molecular systems.

chemical perspective, a resonance contributor is presented as a wavefunction. A linear combination of these resonance contributors (wavefunctions) describes the wavefunction of the molecule. In practice, the molecule can be understood as a resonance hybrid of its various resonance forms. The resonance contributors of a molecule therefore differ only in the assignment of electrons, but the nuclear positions shall remain unchanged. Resonance contributors can be of different significance. The contribution is weighted according to the stability of the Lewis structure. The more stable the Lewis structure, the greater the contribution it makes. A stable Lewis structure should meet the following requirements: (i) maximum number of bonds, (ii) minimum separation of opposite charges, and (iii) showing charge distribution consistent with relative electronegativity.

1.3.3 Aromaticity and Hückel Molecular Orbital (HMO) Theory

Benzene and benzene-like molecules exhibit the property called **aromaticity** [11]. Originally, aromaticity was associated with the unusual reactivity of benzene toward substitution rather than addition reactions. As described by its resonance scheme, benzene can be represented by two equivalent Kekulé structures. Structurally, benzene takes a perfect hexagon shape in which all the C–C bonds possess the same bond strength and length. Energetically, benzene is more stable than that of a non-delocalized hypothetical cyclohexatriene by about 36 kcal mol^{-1}. The extra stabilization comes from a cyclic resonance energy of benzene due to aromaticity. In the common practice, aromaticity can be described by the **Hückel's rule** as follows. Aromatic compounds usually feature a planar cyclic hydrocarbon structure that is fully π-conjugated. The π-electrons carried by the **aromatic** ring should be a (4n + 2) number, where n is 0 or any integer. If the number of π-electrons is 4n, the system is called **anti-aromatic**. Following this rule, planar cyclic structures with (4n + 2) π-electrons in Figure 1.9 can be categorized as being aromatic. The planar cyclic structures with 4n π-electrons are anti-aromatic and therefore cannot exist as stable species. If a molecular structure significantly deviates from planarity, it is viewed as **non-aromatic** due to the disruption of cyclic π-conjugation over a non-planar ring. For example, if cyclooctatetraene adopts a planar conformation, it would exhibit anti-aromaticity. However, the eight-membered ring of cyclooctane favors a tub-shaped conformation due to the angle strain in the eight-member ring (see Figure 1.9). So, in the stable conformation of cyclooctatetraene, the p orbitals of two adjacent C=C bonds are in a nearly perpendicular orientation. This leads to poor orbital overlap and hence the π-electrons are localized in each C=C bond.

Theoretically, the Hückel's rule was derived from the **Hückel molecular orbital** (HMO) theory [12], which is a simplified but very efficient method to elucidate the molecular orbital properties for π-conjugated systems. In the HMO method, only the interactions of adjacent p atomic orbitals are considered. The π

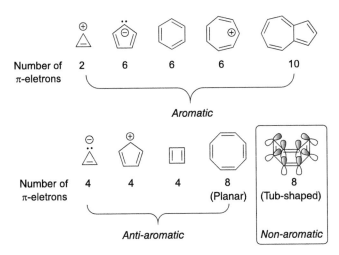

Figure 1.9 Typical hydrocarbon rings showing aromaticity, anti-aromaticity, and non-aromaticity.

MOs resulting from the overlap of a set of p orbitals can be calculated by solving a secular equation. For a linear polyene, the energy levels of HMOs are given by the equation, $E_i = \alpha + m_i\beta$, where α is called the Coulomb integral, β is called the resonance integral, $m_i = 2\cos(j\pi/(n+1))$, and j is the orbital number ($j = 1, 2, 3, ..., n$). For a monocyclic π-conjugated system, the HMO energies are given by the equation, $E_i = \alpha + m_i\beta$, where $m_i = 2\cos(j\pi/n)$. A device called the **Frost circle** can assist to conveniently obtain the MO energies as illustrated in Figure 1.10. First, a perfect polygon is inscribed in a circle with one vertex down. The radius of the circle is 2β, and the level at the center of the circle is set as α. The HMO energy levels are at the positions of the vertexes of the polygon, which can be determined geometrically.

The HMO theory allows the stabilization energy of an aromatic ring such as benzene to be estimated by comparing its total π-electron energy with that of 1,3,5-hexatriene. As shown in Figure 1.11, the two π-conjugated systems have the same number of C=C bonds; however, the lack of cyclic π-conjugation (i.e., aromaticity) makes 1,3,5-hexatriene less stabilized than benzene by 1.012β (ca. 22 kcal mol^{-1}).

Apart from the Hückel's rule, aromaticity can also be quantitatively assessed through methods based on structural, energy, reactivity, and magnetic criteria. The following describes several representative methods that have been popularly used in recent years. Structurally, aromaticity leads to **bond length alternation** (BLA). In this respect, **harmonic oscillator for aromaticity** (HOMA) is popularly used as an index of aromaticity [13]. HOMA can be calculated by the following equation:

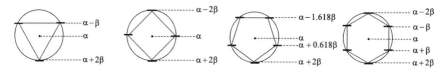

Figure 1.10 Energy level diagrams for monocyclic π-conjugated systems generated with the aid of the Frost circle.

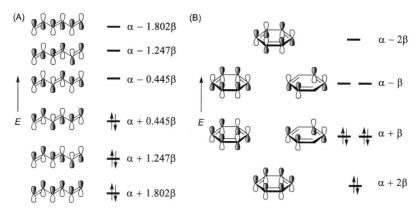

Figure 1.11 HMO plots for 1,3,5-hexatriene (A) and benzene (B).

$$\text{HOMA} = 1 - [a(R_{opt} - R_{av})^2 + a/n\sum(R_{av} - R_i)^2]$$

where R_{av} is the average bond length, R_{opt} is the optimum bond length, R_i is an individual bond length, n is the number of bonds, and α is an empirical constant that makes the HOMA of a localized non-aromatic structure *zero*. Energetically, aromaticity can be assessed by an index called **aromatic stabilization energy** (ASE), which is calculated based on a carefully selected thermodynamic cycle for a hypothetical **homodesmotic reaction** that involves breaking the π-conjugation in the ring of interest. A homodesmotic reaction belongs to the class of so-called **isodesmic reactions**, in which reactants and products contain equal numbers of atoms in corresponding states of hybridization. Moreover, there is matching of the carbon–hydrogen bonds in terms of the number of hydrogen atoms joined to the individual carbon atoms. They serve as useful tools for studying the energy and thermochemistry of molecules.

According to the HMO theory, aromatic systems tend to possess a relatively large energy gap between the **highest occupied molecular orbital** (HOMO) and **lowest unoccupied molecular orbital** (LUMO). According to the Koopmans theorem, the HOMO-LUMO gap ($\Delta E_{\text{HOMO-LUMO}}$) is equal to the **hardness** (η) of the molecule, which is a quantity introduced by Pearsons in his well-known

hard/soft acids and bases (**HSAB**) principle [14, 15]. The relatively large $\Delta E_{\text{HOMO-LUMO}}$ of an aromatic compound (e.g., benzene) is consistent with its reduced reactivity toward electrophiles. In contrast, anti-aromatic compounds have small HOMO-LUMO gaps and therefore are reactive and relatively unstable.

Nuclear magnetic resonance (NMR) spectroscopy provides an effective tool to assess aromaticity, since the delocalized π-electrons in an aromatic compound tend to form a diamagnetic ring current when subjected to the influence of an external magnetic field [16]. The ring current strongly affects the ^{1}H and ^{13}C chemical shifts. Schleyer and co-workers devised a popular method to correlate chemical shift (δ) and aromaticity, called the **nucleus independent chemical shift** (NICS) [17]. It evaluates aromaticity based on the magnetic shielding at the center of a ring of interest calculated using *ab initio* or density functional theory (DFT) methods. When calculating NICS values, dummy atoms are placed at the centroid of an aromatic ring and various positions above it. A dummy atom is a hypothetic atom that does not possess any electrons or a nucleus. It is therefore only used to define a coordinate for probing the chemical shielding at its location in quantum mechanics calculations. The negative value of the calculated chemical shift (δ) of the dummy atom placed at the ring center gives the NICS(0) (Figure 1.12). A ring that shows a large negative NICS(0) value is considered as aromatic, while a large positive value indicates anti-aromaticity. For example, benzene and benzenoid hydrocarbons show NICS(0) values of -9 to -10 ppm. Anti-aromatic cyclobutadiene features a NICS(0) value of $+27.6$ ppm. In addition to the ring center, the dummy atom can also be placed at a position 1.0 Å vertically above the center of the ring to obtain the NICS(1.0) value. The advantage of using NICS(1.0) as a criterion for aromaticity is that the ring current usually maximizes at this position and the influences of other structural factors are relatively small.

Aromaticity can be found in cyclic π-conjugated systems, where the π-conjugation of the ring is interrupted by one or more than one saturated linkage(s) but orbital overlap still occurs due to the geometry of the molecule. Such structures are deemed to possess **homoaromaticity** [18]. Furthermore, the systems can be defined as mono-, bis-, and tris-homoaromatic according to the number of interruptions to

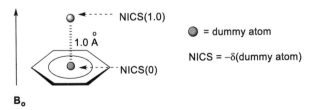

Figure 1.12 Calculations of NICS(0) and NICS(1.0) for benzene with the aid of dummy atoms (shaded circles). B_0 is the external magnetic field and δ is the calculated chemical shift (in ppm).

π-delocalization. Figure 1.13 lists a selection of mono-homoaromatic cationic and anionic cyclic π-systems, where (4n + 2) π-electrons are delocalized through a loop of overlapped p orbitals. These homoaromatic ions have been found to show enhanced stability; however, the stabilizing effects are not as strong as those in the aromatic systems.

The delocalization of a charge in the cationic and anionic systems in Figure 1.13 is a significant factor enabling them to exhibit homoaromaticity. Examples of neutral homoaromatic systems have also been proposed and studied, but the homoaromaticity of many of such systems has been determined to be either very weak or in dispute. For example, Figure 1.14 lists a series of neutral systems showing mono-, bis-, and tris-homoaromaticity. Cycloheptatriene (Figure 1.14A) has a saturated CH_2 group linking the two ends of a conjugated triene. Partial π-orbital overlap between the two ends was proposed to form a benzene-like delocalization pattern. It has been regarded as a marginally homoaromatic system. Semibullvalene (Figure 1.14B) can undergo a degenerate Cope rearrangement through a low-lying

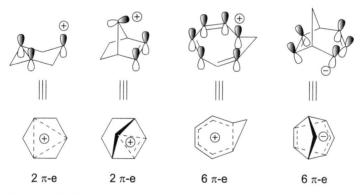

| 2 π-e | 2 π-e | 6 π-e | 6 π-e |

Figure 1.13 Examples of cationic and anionic homoaromatic systems.

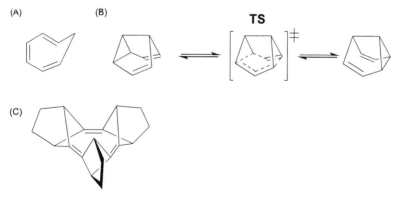

Figure 1.14 Examples of neutral homoaromatic systems.

transition state ($\Delta G^{\neq} = 6.2$ kcal mol^{-1}). In this transition state, four π-electrons and two σ-electrons are fully delocalized, exhibiting a bis-homoaromaticity. Stahl et al. investigated a strained cyclic triene system (Figure 1.14C) through systematic computational analyses and convincingly demonstrated it as a neutral tris-homoaromatic system [19].

Fully π-conjugated cyclic systems may feature a Möbius topology in terms of the arrangement of p-orbital overlap to show aromaticity, which is known as the **Möbius aromaticity** [20]. A Möbius strip can be constructed from a rectangular strip by first giving one of the ends a one-half twist and then affixing the two ends together. This type of strip exhibits interesting properties, such as having only one side and remaining in one piece when split down the middle. Figure 1.15 depicts the contrast between typical planar cyclic π-conjugation (i.e., Hückel system) and a Möbius π-system. In contrast to the Hückel system, a Möbius ring structure carrying 4n π-electrons is considered as aromatic, while (4n + 2) π-electrons lead to anti-aromaticity. Möbius aromatic compounds do not naturally exist; however, synthetic chemists have successfully constructed certain twisted annulenes and porphyrins that show Möbius aromaticity [21–23].

1.3.4 Hyperconjugation and Orbital Interactions

Resonance theory explains the properties of electrons delocalizing among π-bonds and/or non-bonding p orbitals that show a significant degree of overlap. Electron pairs of σ-bonds are not considered migratable under the assumption that they do not overlap significantly with π and p orbitals. Therefore, σ-bonds are not allowed to be broken in resonance scheme drawing. Nevertheless, electron delocalization between a σ–bond and its neighboring π/p orbitals can still occur, resulting in a significant effect on chemical stability and reactivity. Such effect is known as the **hyperconjugation effect** [24], which can be conveniently explained by the **orbital interaction theory** [25–27]. Figure 1.16 illustrates the stabilizing effect on the ethyl cation and propene due to hyperconjugation. In the ethyl cation, the C–H σ-bond overlaps with the empty p orbital of the carbocation to some extent,

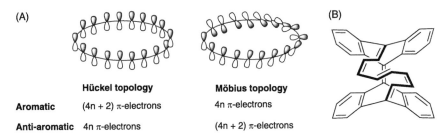

(A)

	Hückel topology	Möbius topology
Aromatic	(4n + 2) π-electrons	4n π-electrons
Anti-aromatic	4n π-electrons	(4n + 2) π-electrons

(B)

Figure 1.15 (A) Illustrations of π-systems showing Hückel and Möbius topologies. (B) A stable Möbius annulene.

although the overlap is not as large as the overlap of two p orbitals aligned in parallel. The delocalization of C–H bonding electrons to the p orbital can be illustrated by using a resonance scheme, which reveals that the hydrogen atom of the C–H has a partial cation character and the C–C bond has a partial double bond character. Qualitative molecular orbital drawings can also assist in understanding the stabilizing effect of hyperconjugation. According to the molecular orbital theory, the mixing of two overlapping orbitals results in two molecular orbitals. The low-lying orbital is called a **bonding orbital**, and the high-lying orbital is called an **antibonding orbital**. As shown in Figure 1.16, the interaction of the C–H bond and the adjacent empty p orbital leads to an electron pair populated in the low-lying bonding orbital, while the high-lying antibonding orbital remains unoccupied. Overall, the system gains stability from the hyperconjugation effect. Similarly, the stabilizing effect of an alkyl group substituted on an alkene unit can be explained by hyperconjugation. Take propene as an example, hyperconjugation occurs between the methyl C–H bond and the antibonding orbital of the C=C double bond. The resonance scheme drawing reveals that one of the C=C carbon atoms possesses a partial carbanion character, indicating that the π-electron density is enhanced by alkyl substitution.

Stabilizing effects similar to hyperconjugation can be found in systems where a donor orbital effectively interacts with an acceptor orbital. A donor orbital can be an electron-rich – bond or a non-bonding orbital occupied by an electron lone pair. An acceptor orbital is usually an empty p orbital, the antibonding orbital of a polar C–X bond (σ^*), or a low-lying antibonding π^* orbital. For example, the C–H bond (donor) in 1,2-difluoroethane interacts with the antibonding orbital of the adjacent C–F bond (acceptor) when they are aligned antiparallel to gain stabilization (Figure 1.17A). As such, the 1,2-difluoroethane molecule prefers a conformation where the two fluoro groups are gauched, known as the **gauche effect**. When a filled p or π orbital interacts with an adjacent σ^* orbital, stabilization is gained. This is generally referred to as the **negative hyperconjugation** [27, 28]. For example, the methoxy group of 2-methoxytetrahydropyran prefers to take the axial position rather than the equatorial position (Figure 1.17B). This phenomenon is known as the **anomeric effect**, the origin of which can be rationalized by the maximum orbital interaction between one of the oxygen lone pair orbitals and the vertically oriented C–O antibonding orbital (σ^*_{C-O}) in this conformer. Negative hyperconjugation can lead to significantly altered spectroscopic properties. For example, the stretching of the C–H bond of an aldehyde occurs at 2700–2800 cm^{-1}, which is much lower than the vibrational frequency of a typical sp^2 C–H bond (around 3050 cm^{-1}). This result can be explained by the fact that the one of the oxygen lone pair orbitals is aligned antiparallel to the C–H bond, enhancing the interaction between the oxygen lone pair orbital and the C–H antibonding orbital. As such, the C–H bond of an aldehyde is lengthened and weakened. This effect can be illustrated by the negative hyperconjugation scheme in Figure 1.17C.

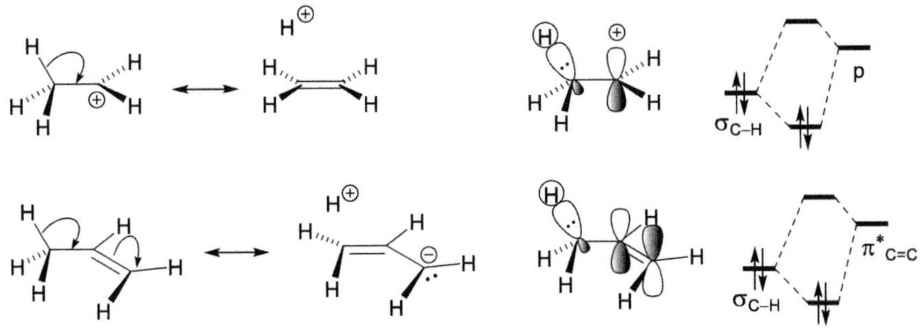

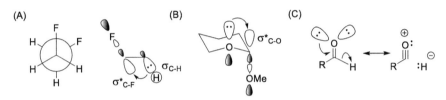

Figure 1.16 Hyperconjugation effects taking place in ethyl cation and propene.

Figure 1.17 Examples of hyperconjugation interactions in organic structures.

1.4 Non-Covalent Interactions and Supramolecular Chemistry

Many organic nanostructures and materials are assembled at the supramolecular level, utilizing various non-covalent interactions as the key forces to influence and control the structural and electronic properties as well as material functions of interest. **Supramolecular chemistry**, as defined by Jean-Marie Lehn, deals with the "chemistry beyond molecules." The original studies of supramolecular chemistry began with the development of guest–host chemistry, mainly focusing on understanding the non-covalent interactions between a host and a guest molecule. Nowadays this field has been vastly expanded to cover many disciplines. Non-covalent interactions arise from various attractive and repulsive forces between molecules and ions. They can be generally classified into the following types.

1.4.1 Electrostatic Interactions Involving Ions and Dipoles

Ions of opposite charges show strong electrostatic (i.e., Coulombic) attraction. The ion–ion interactions are usually of comparable strength to covalent bonding, leading to the formation of ionic solids, such as various inorganic salts. Some specially designed organic cations can attract anions to form supramolecular assemblies, in which the ion–ion attraction plays an important role. For example, the organic trication derived from 1,3,5-tris(bromomethyl)-2,4,6-trimethylbenzene and (1,4-diazabicyclo[2.2.2]octane) (DABCO) can form a crystalline material through ion–ion interactions with ferricyanide anion (Figure 1.18) [29].

An ion can interact with a molecule that possesses a permanent dipole. Such an interaction is called an **ion–dipole interaction**, the strength of which is dependent on the geometric parameters for the arrangement of the ion and the dipole. As shown in Figure 1.19A, the dipole moment (μ) can be determined by the equation, $\mu = q_1 \times r$, where q_1 is the amount of charge accumulated at the end of dipole and r is the distance of charge separation. The potential energy (E) for the dipole interacting with an ion carrying a charge of q_2 can thus be determined using the following equation

$$E = \frac{\mu q_2 \cos\theta}{4\pi\varepsilon\varepsilon_0 d^2} \tag{1.1}$$

where θ and d are the angle and distance of the ion with respect to the dipole (see Figure 1.19A), ε is the relative permittivity (or dielectric constant) of the medium, and ε_0 is the vacuum permittivity.

Ion–dipole interaction plays a critical role in the solvation of cations by strongly dipolar solvents, such as water, DMF, and DMSO. As illustrated in Figure 1.19B,

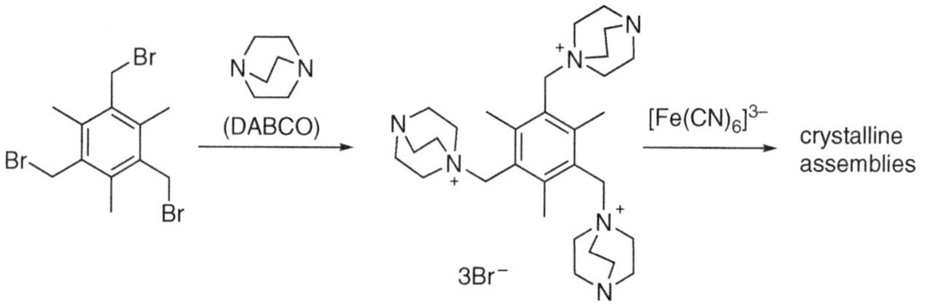

Figure 1.18 A designed organic trication for self-assembly with ferricyanide.

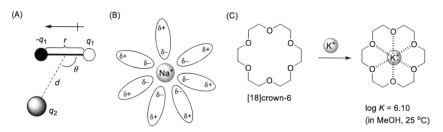

Figure 1.19 (A) Parameters of an ion–dipole interaction. (B) Favored arrangement of dipolar solvent molecules surrounding a sodium cation. (C) Complexation of [18]crown-6 with a potassium cation.

polar solvent molecules can form a solvation shell in which the partially negatively charged ends of the solvent molecules are drawn close to the cation through
ion–dipole interactions. For cations with the same charge but different sizes, the
ion–dipole interaction energy increases with decreasing ion size. This is because
the smaller radius of the ion, the higher the charge density per unit area it affords
and hence the resulting Columbic interaction. Cyclic and cage-like organic molecules with preorganized polar bonds can act as good supramolecular hosts for various cations. A well-known example is the class of crown ether molecules. For
example, [18]crown-6 (IUPAC name 1,4,7,10,13,16-hexaoxacyclooctadecane) can
form a stable 1:1 complex with a potassium ion. In this complex, all the polar C–O
bonds are favorably arranged to enhance the ion–dipole attractions (Figure 1.19C).

Similar to an ion–dipole interaction, two dipoles in close proximity can give rise
to attractive or repulsive forces (**dipole–dipole interaction**) depending on their
relative orientations. As shown in Figure 1.20A, when two dipoles are closely
positioned, their ends with opposite charges would induce Coulombic attraction,
while the ends whose charges have the same sign repel each other. The potential
energy of a dipole–dipole interaction can be calculated following the scheme and
equation illustrated in Figure 1.20B. Since the energy of dipole–dipole interaction
is inversely proportional to the cube of their distance (d^3) apart, the strength of a
dipole–dipole force is very sensitive to distance. Compared with ion–dipole interactions, the strength of dipole–dipole interactions is weaker and more dependent
on direction. Usually, polar molecules prefer to be aligned in such a way that the
positively charged pole closely approaches the negatively charged pole (see Figure
1.20C) to give the strongest possible interaction. For molecules with resonance-
stabilized charge separation, dipole–dipole interactions among them are further
enhanced. For example, amides show strong resonance effects in which the lone

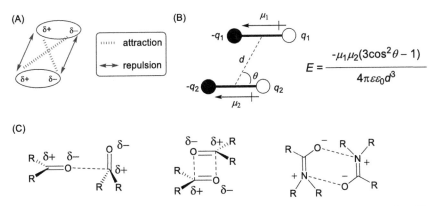

Figure 1.20 (A) Electrostatic attraction and repulsion in a dipole–dipole interaction. (B)
Calculations of dipole–dipole interaction energy. (C) Dipole–dipole interactions among
ketone and amide molecules.

pair electrons of the N atom are delocalized with the antibonding orbital of the adjacent $C=O$ π-bond. As such, an amide molecule would show a large permanent dipole with enhanced charges on the N and O ends, which in turn strengthens the pairing of two oppositely aligned amide molecules through dipole–dipole interactions (Figure 1.20C). The strong intermolecular attractions thus make amides exhibit much higher boiling and melting points than other carboxylic acid derivatives with similar molecular sizes.

Furthermore, it is worth noting that when two dipoles are aligned at an angle (θ) equal to 54.7°, the term ($3\cos^2\theta-1$) turns into zero and the dipole–dipole interaction vanishes. This angle is known as the **magic angle**, which is used in a special technique for solid-state NMR spectroscopy, namely **magic angle spinning** (MAS), to eliminate the line broadening effect due to nuclear dipolar interactions.

1.4.2 Hydrogen Bonding Interactions

Hydrogen bonding is a very important type of binding force, specifically referred to the attractive interaction of a hydrogen atom with an electronegative atom [30]. Usually, a hydrogen bond exists in the form of D–H$\cdots$A, where D–H is a **hydrogen bond donor** and A is a **hydrogen bond acceptor**. The D unit in D–H is usually an electronegative atom such as N, O, or S, so that the hydrogen atom connected to it exhibits a partial positive charge and becomes somewhat acidic. The A group often bears a relatively high-energy electron lone pair(s) and possesses a partial negative charge, making it attractive to the acidic proton of D–H through Coulombic interaction. In this sense, hydrogen bonding can be viewed as a special type of dipole–dipole interaction or acid–base interaction.

Hydrogen bonding ubiquitously exists in supramolecular chemistry and is regarded as the "master key to molecular recognition." Normal hydrogen bonds show bond strengths ranging from about 1 to 14 kcal mol^{-1}. The strength of a hydrogen bond is associated with the electronegativity of the two counterparts, D in D–H and A. For a hydrogen bond donor (D–H), its hydrogen bond donating ability increases with increasing electronegativity of the D unit. So is the hydrogen bond accepting ability for an A group. When a hydrogen bond involves an ion (e.g., hydronium ion or fluoride anion), the hydrogen bond strength can be considerably enhanced (see Figure 1.21).

Hydrogen-bonded systems can adopt different geometries as summarized in Figure 1.22. These geometries result from direct interactions of hydrogen bond donor and acceptor groups, and they are referred to as the **primary hydrogen bonding interactions**. In a D–H$\cdots$A interaction, the hydrogen bond angle can be linear or deviated from linearity (bent), depending on the orientation of the lone pair electrons on the A group. For example, when the A group involves only one electron lone pair (e.g., amino, cyano), a colinear arrangement of D–H$\cdots$A is expected. When there are more than one electron lone pairs on the A group

Common hydrogen bonds

$S-H\cdots S-H$

1.7 kcal mol^{-1}

$N-H\cdots N$

4.1 kcal mol^{-1}

$O-H\cdots O-H$

6.9 kcal mol^{-1}

$O-H\cdots Cl^{\ominus}$

13.1 kcal mol^{-1}

Strong hydrogen bonds

$O-H\cdots F^{\ominus}$

23.4 kcal mol^{-1}

$O-H\cdots O^{\oplus}$

36.0 kcal mol^{-1}

$F-H\cdots F^{\ominus}$

40.3 kcal mol^{-1}

Figure 1.21 Examples of hydrogen-bonding interactions and hydrogen bond strengths.

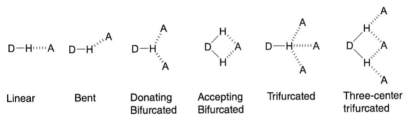

| Linear | Bent | Donating Bifurcated | Accepting Bifurcated | Trifurcated | Three-center trifurcated |

Figure 1.22 Geometries of primary hydrogen bonding interactions.

(e.g., hydroxy, carbonyl), the hydrogen-bonded geometry is consistent with the electron lone pair orientations, which can be reasonably predicted by the VSEPR theory. Usually, the linear hydrogen-bonding motifs show stronger binding forces than the bent ones.

When hydrogen bonds are formed between two or more molecules of the same type, they are called **homo-intermolecular hydrogen bonds**. For example, imidazole molecules can form a linear network through homo-intermolecular hydrogen bonds (Figure 1.23A). A carboxylic acid and an amide (primary and secondary) can dimerize through homo-intermolecular double hydrogen bonds (Figure 1.23A). Hydrogen bonding interactions between molecules of different types give the **hetero-intermolecular hydrogen bonds** (Figure 1.23B). A hydrogen bond can also occur within the same molecule, known as an **intramolecular hydrogen bond** (Figure 1.23C). Usually, an intramolecular hydrogen bond favors the formation of a stable six-member ring motif. On the other hand, intramolecular hydrogen-bonded five-, seven-, and eight-member rings also exist.

When hydrogen bonding interactions occur between arrays of hydrogen bond donors and acceptors which are in close proximity, **secondary hydrogen bonding interactions** also play a notable role in the stability of the hydrogen-bonded assemblies. Figure 1.24 exemplifies three hydrogen-bonded molecular

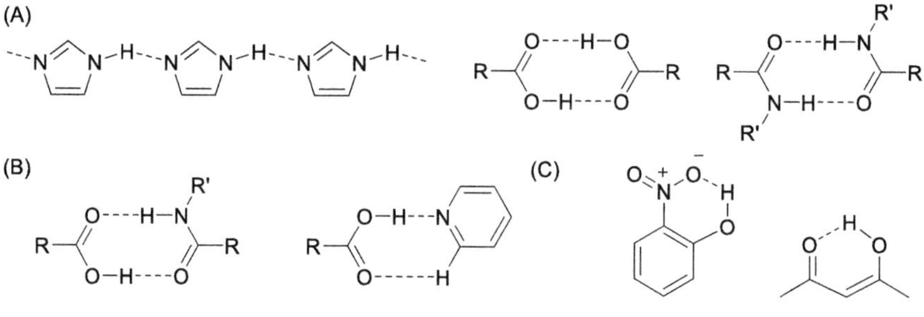

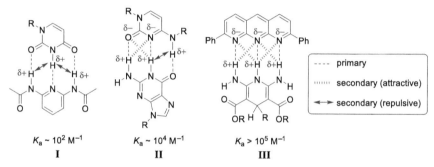

Figure 1.23 Examples of (A) homo-intermolecular, (B) hetero-intermolecular, and (C) intramolecular hydrogen bonds.

$K_a \sim 10^2\ M^{-1}$ $K_a \sim 10^4\ M^{-1}$ $K_a > 10^5\ M^{-1}$

I   II   III

---- primary

········ secondary (attractive)

◄──► secondary (repulsive)

Figure 1.24 Hydrogen-bonded supramolecular assemblies with varied degrees of secondary hydrogen bonding interactions.

pairs, each of which consists of three hydrogen bonds. The hydrogen bond donor (D) and acceptor (A) groups in these structures are in different arrangements. The first hydrogen-bonded assembly **I** contain two counterparts, which are in the DAD and ADA motif, respectively. The secondary interactions as illustrated afford only repulsive forces. In assembly **II**, the motifs are changed to DDA and AAD. As such, the secondary interactions give two attractive forces and one repulsive force. The third structure **III** is a DDD and AAA assembly, where all the secondary interactions give attractive forces. The differences in their secondary hydrogen bonding interactions make these three hydrogen-bonded assemblies show very different association constants (K_a). Of these structures, only **III** exhibits all attractive interactions, while **I** and **II** show repulsive secondary interactions to varied extents. As a result, **III** possesses the highest stability among them.

The formation and properties of hydrogen bonds can be investigated by a variety of experimental methods, such as vibrational and NMR spectroscopic analyses. For example, the hydrogen bonding interaction in a D–H···A system would lengthen the D–H bond and give rise to new vibrational modes due to the H···A bond. In the NMR analysis, pronounced proton deshielding for the

H in D–H can be observed as a result of hydrogen bonding. Energetically, the formation of a hydrogen-bonded assembly makes the Gibbs energy of formation greater than the thermal energy of the system, which can be experimentally determined. Hydrogen bonds are directional and can strongly influence crystal packing modes. Crystallographic analysis can therefore cast a deep insight into the roles of hydrogen bonds in the solid state. Besides experimental methods, computational studies based on quantum chemical theories are of great value in characterizing and assessing hydrogen bonds. A popular theoretical method, which is often used in combination with experimental analysis, is to perform critical point analysis on the electron density topology of the molecular system based on the theory of **atoms in molecules** (AIM) proposed by Bader [31]. In 2011, a task group of International Union of Pure and Applied Chemistry (IUPAC) published an article, entitled *Definition of the hydrogen bond (IUPAC Recommendations 2011)*, which offers updated definition and criteria for the hydrogen bond [32].

1.4.3 Interactions Involving π-Systems

As discussed above, π-electrons delocalize on the skeleton of an aromatic ring, resulting in an extra stabilization attributed to aromaticity. The electronic properties of aromatic compounds would also lead to unique π-effects in their non-covalent interactions, commonly called **π-stacking** or **π–π interactions** [33–35]. In benzene, for example, high electron density is present around the central region of the benzene ring, while the six hydrogen atoms on the edges are electron deficient. Such an electron distribution pattern can be rationalized by the polar nature of benzene's sp^2 C–H bonds. A typical C–H bond in an alkane is non-polar, since the sp^3 hybridized carbon has similar electronegativity to hydrogen. The sp^2 hybridized carbons in benzene, on the other hand, are more electronegative due to their larger percentage of s character (33.3%) involved in hybridization. As a result, the C–H bonds of benzene show a significant bond dipole moment. Although the hexagon-shape of benzene makes the C–H bond dipoles cancel one another to yield a zero net molecular dipole moment, a quadrupole moment exists in benzene as illustrated in Figure 1.25. Because of this, benzene can still show electrostatic attraction to various cationic and polar species in a manner similar to ion–dipole and dipole–dipole interactions.

Figure 1.26 shows various modes of π-interactions for benzene. Since benzene has a significant quadrupole moment, it can interact with a cation through electrostatic attraction. Figure 1.26A shows the equilibrium geometry of a benzene/cation complex, known as a cation-π interaction. Herein the cation favors alignment with the center of benzene, which is the orientation that maximizes the electrostatic attraction. The equilibrium distances and binding energies for the 1:1 complexes of benzene with various cations are listed in Table 1.1. As

(A) (B)

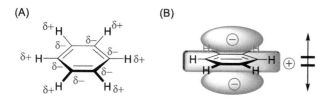

Figure 1.25 Illustrations of (A) the polar C–H bonds and (B) quadrupoles of benzene.

(A) (B) (C) (D) (E)

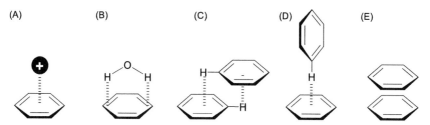

Figure 1.26 Various π-interactions of benzene. (A) cation–π interaction, (B) polar-π interaction, (C) parallel-displaced π-stacking, (D) edge-to-face π-stacking, (E) eclipsed face-to-face π-stacking.

Table 1.1 Equilibrium distances (D) and binding energies ($E_{binding}$) for 1:1 complexes of benzene with various alkali and alkaline metal cations [36].

Cation	D (Å)	$E_{binding}$ (kcal mol^{-1})
Li$^+$	1.85	−35.9
Na$^+$	2.44	−21.5
K$^+$	3.05	−13.1
Mg^{2+}	1.95	−111.9
Ca^{2+}	2.47	−66.7

Data calculated at the MP2/Sadlej level of theory.

can be seen, the binding strength increases with decreasing ion size and increasing charge. Besides metal cations, benzene and benzene-containing aromatic compounds can also interact with organic cations (e.g., ammonium ions) through the cation–π interaction.

Benzene can also show attractive interactions with molecules with strong permanent dipoles, which are called **polar-π interactions**. For example, the interaction energy of benzene and water was experimentally determined to be about 2.4 kcal mol^{-1}. The binding mode is depicted in Figure 1.26B, in which the two partially positively charged protons of water are in contact with the electron-rich sp^2 carbons of benzene. In a sense, this interaction is a weak hydrogen bonding interaction. It is worth noting that benzene is a hydrophobic compound that is immiscible with water. However, the hydrophobic (water-hating) behavior is not a result of repulsive interactions between benzene and water molecules. We shall be very clear that benzene and water attract one another through polar–π interactions. The hydrophobic behavior mainly comes from an entropic origin which will be discussed in detail in Section 1.4.6.

π-Stacking can occur among molecules of benzene and benzene-like aromatic molecules. Two commonly seen stacking modes of benzene are illustrated in Figure 1.26C and D, which are called parallel-displaced (slipped) and edge-to-face (T-shaped) π–stacking, respectively. In these two stacking modes, the positively charged edge of one benzene molecule is in a close contact with the centroid region of another benzene where the negatively charge electron cloud is populated. Therefore, the parallel-displaced and edge-to-face stackings are both energetically favored. The edge-to-face stacking is energetically favored and therefore more frequently observed in crystal packing structures. The eclipsed face-to-face (sandwich) stacking of benzene (Figure 1.26E) align the regions of positive and negative electrostatic potentials, respectively. Actually, this stacking mode leads to an electrostatic repulsion and hence is disfavored. It is worth cautioning that the term "π-stacking" or "π–π interactions" would mislead one to consider that the eclipsed face-to-face is favored. So, a clear understanding of the origins of these π–π interactions is necessary before using such terms.

1.4.4 Induced-Dipole Forces

Molecules that are lacking permanent dipoles can still show attractive electrostatic interaction with a polar species or a nonpolar molecule in close proximity [37]. When subjected to an external field, a molecule will respond with changes in the distribution of its electron density, which can be described by a property called **polarizability** (α). The polarization of the electron cloud of a nonpolar molecule creates a temporary dipole moment (μ), which in turn gives electrostatic attraction toward ions or molecules nearby. Figure 1.27 illustrates three exemplar scenarios of induced-dipole formation.

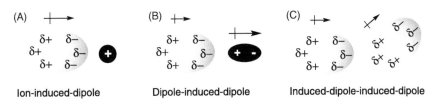

Ion-induced-dipole   Dipole-induced-dipole   Induced-dipole-induced-dipole

Figure 1.27 Schematic illustrations of the formation of induced-dipoles of nonpolar molecules.

Figure 1.27A depicts the ion-induced-dipole interaction, the energy of which (E) can be determined by the following equation:

$$E = \frac{-q^2\alpha}{(4\pi\varepsilon\varepsilon_0)^2 r^4} \qquad (1.2)$$

where q is the charge of the cation, α is the polarizability of the molecule, and r is the distance between the centroids of the ion and the molecule. As can be seen that the interaction energy E is dependent on $1/r^4$, indicating that the attractive force would diminish very rapidly as the distance increases. Likewise, a molecule with a permanent dipole moment can polarize a nonpolar molecule when approaches it closely. The induced dipole of the nonpolar molecule aligns with the permanent dipole to give an electrostatic attraction, known as an **dipole-induced-dipole interaction** (Figure 1.27B). Compared with the ion-induced-dipole interaction, this force is even more sensitive to distance, showing a dependence on $1/r^6$.

When a nonpolar molecule is instantaneously polarized to generate a temporary dipole, it can exert the same effect on a nearby nonpolar molecular as a permanently dipolar molecule does (see Figure 1.27C). This type of force is called an **induced-dipole-induced-dipole interaction**, which has also been well known as a **London dispersion interaction**. The magnitude of such interaction is considerably small, but it plays a significant role in the intermolecular interactions in condensed liquids such as alkanes and alkenes. For example, an alkane with a linear zigzag molecular backbone shows a higher boiling point than its structural isomers that show branched molecular structures. This is because the linear alkane molecules can be more efficiently packed (due to more effective contact areas) to afford stronger intermolecular attractions.

1.4.5 Charge-Transfer Interactions

When a donor (D) and an acceptor (A) molecule are closely stacked, a fraction of electronic charge is transferred from the HOMO of the donor to the LUMO of the acceptor, resulting in an electrostatic attraction between them, called the **charge-transfer (CT) interaction**. Like hydrogen bonding, the CT interaction is a directional intermolecular force that plays an important role in supramolecular chemistry. A vast array of redox-active organic π-chromophores, mainly aromatic compounds, can form D$\cdots$A complexes through CT interactions. Figure 1.28 shows a selection of organic π-donors and π-acceptors [38].

The strength of D$\cdots$A complexation is dependent on several of parameters, including the donating and accepting ability of the D and A molecules, steric crowding in their structures, the shape complementarity of their π-surfaces, and

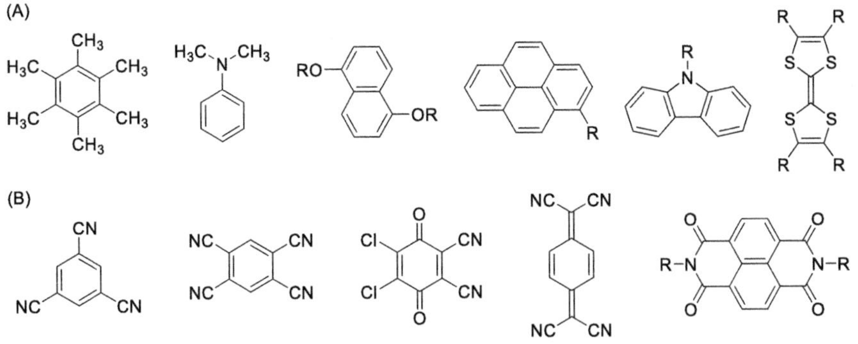

(A)

(B)

Figure 1.28 Examples of (A) organic π-donors and (B) organic π-acceptors.

the solvent effects. The binding affinity between neutral π-donor and π-acceptor molecules in the solution phase is usually not very strong. For example, hexamethylbenzene (HMB) is an organic π-donor that can form a 1:1 complex with electron-deficient 1,3,5-trinitrobenzene (TNB), affording an association constant (K_a) of 15.5 ± 0.4 M^{-1} in cyclohexane at 20 °C. When the solvent is switched to chloroform, the K_a is attenuated to 0.76 ± 0.05 M^{-1}. Tetrathiafulvalene (TTF) and its derivatives are an important class of organic π-donors. Their excellent electron-donating ability is due to their aromaticity-stabilized oxidized states. As shown in Figure 1.29A, a TTF molecule can sequentially release two electrons to yield a radical cation and dication, respectively. In each of the cationic products, the five-membered dithiolium ring features a 6 π-electron cyclic conjugation pattern. According to the Hückel's rule, the dithiolium ring is aromatic. The aromaticity-stabilized oxidized products of TTF enable it to release electrons at relatively low potentials. TTF can efficiently complex with cyclobis(paraquat-*p*-phenylene) (**CBPQT^{4+}**) with a large association constant in solution (Figure 1.29B). Such CT interactions have been extensively used in the formation of interlocked compounds such as rotaxanes and catenanes, to which Sir J. Fraser Stoddart has made tremendous contributions, leading him to winning the 2016 Nobel Prize in Chemistry.

In the solid state, TTF and its derivatives can form CT salts with strong π-acceptors, such as 7,7,8,8-tetracyano-*p*-quinodimethane (TCNQ). Figure 1.30 illustrates the solid-state structure of a CT salt of dibenzo-tetrathiafulvalene (DB-TTF) and TCNQ, in which the D and A molecules are organized to form [···D–A–D–A···] alternating stacks through CT interactions [39]. TTF-based CT salts show intriguing metallic and semiconducting behavior useful for organic electronic applications. Besides crystalline solids, CT interactions can also be utilized as a major driving force to generate functional materials such as organogels, foldamers, and liquid crystals.

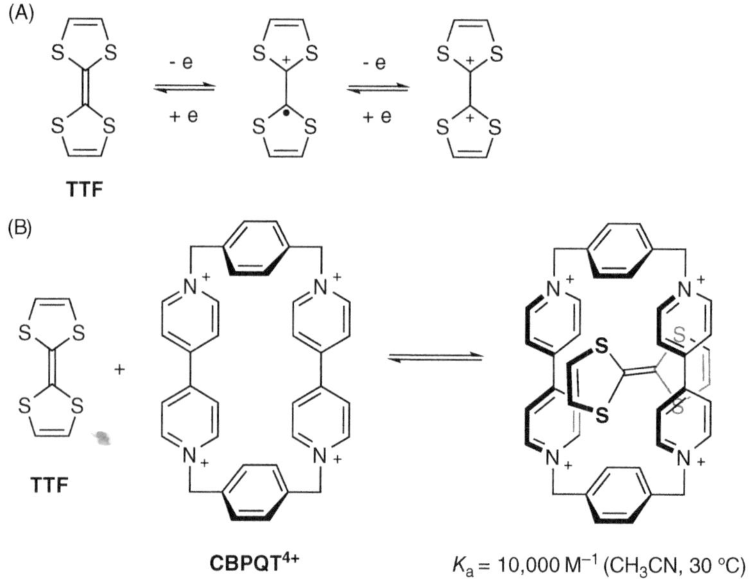

(A)

TTF

(B)

TTF

+

CBPQT^{4+}

$K_a = 10,000\ M^{-1}$ (CH$_3$CN, 30 °C)

Figure 1.29 (A) Stepwise oxidation of tetrathiafulvalene (TTF). (B) 1:1 Complexation of TTF and CBPQT^{4+} ion in the solution phase.

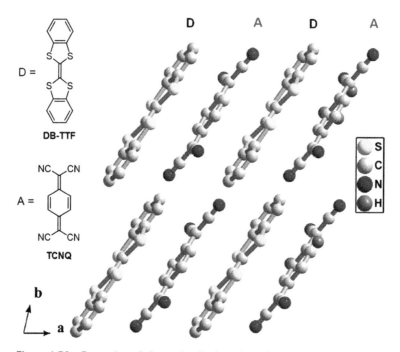

D =

DB-TTF

A =

TCNQ

Figure 1.30 Formation of alternating D–A stacks of DB-TTF and TCNQ. (Reproduced from *CrystEngComm* 2014, *16*, 5968–5983).

1.4.6 Hydrophobic Effects

Hydrophobic effects play an important role in chemistry and biology. The term **hydrophobic** is derived from *hydro* (water) and *phobos* (fear), describing the tendency of nonpolar organic compounds (e.g., aliphatic hydrocarbons) toward aggregation when dispersed in water. The reason for these compounds to show **hydrophobicity** is not because they somewhat "fear" or repulsively interact with water molecules. Instead, hydrophobicity is a result of the differences between solvating water molecules and the bulk water molecules. In a solvation process, two energetic components come into play. The enthalpic (ΔH) effect describes the solvent/solute interactions, such as ion–dipole and dipole–dipole attractions. The entropic (ΔS) effect reflects the changes in the degree of ordering of solvent/solute molecules. As schematically illustrated in Figure 1.31, when individual molecules of a nonpolar compound (e.g., benzene) are dissolved in water, the molecular surface

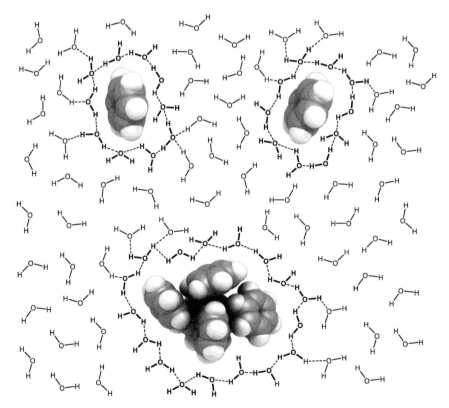

Figure 1.31 Water solvation of individual molecules and aggregates of benzene. Solvating water molecules (highlighted in bold) create cavities to host benzene molecules (solutes).

of each benzene molecule is surrounded by a cluster of water molecules to form a solvation shell, which in turn creates a cavity hosting the solute molecule. Water molecules in the solvation shell are organized in a rigid, quasi-crystalline structure, losing certain degrees of rotational and translational freedom. Therefore, the water molecules in the solvation shell are arranged in greater order than bulky water molecules. As a result, the total entropy of water as the solvent is decreased relative to its entropy before solvation. When benzene molecules approach one other to form aggregates, their effective surface area in contact with water is reduced, liberating some water molecules from the solvation shells. In other words, the number of ordered solvating water molecules per benzene is lowered, and consequently the entropy of water as the solvent is increased after aggregation of benzene. As discussed in Section 1.4.3 the water–benzene interaction is attractive with a strength equivalent to weak hydrogen bonding (i.e., the $\Delta H_{\text{solvent-solute}}$ term is a negative value). Nevertheless, the increase in the entropy term dominates the thermodynamics of benzene dispersion in water. The effect of the $T\Delta S_{\text{solvent}}$ term for non-aggregated benzene molecules is larger than the $\Delta H_{\text{solvent-solute}}$ term (ΔG positive) but smaller for aggregated benzene molecules (ΔG negative). Therefore, the aggregation of benzene molecules in water is a thermodynamically favored process.

Hydrophobicity can serve as an important driving force in guest–host chemistry. Figure 1.32 shows a class of compounds named cucurbit[n]urils, which are macrocycles of glycoluril linked through methylene linkages [40]. The nomenclature of cucurbiturils is derived from their resemblance to a pumpkin of the family of *Cucurbitaceae*. Cucurbit[n]urils are highly polar organic compounds with good water solubility. They possess a sufficiently large inner cavity to act as a molecular container to host a variety of neutral and cationic guest species in water. For example, cyclohexane can be encapsulated inside cucurbit[6]uril with an association constant (K_a) of 1300 M^{-1} in water at pH 3. Usually, the binding of a cucurbituril with a neutral molecule is driven by the hydrophobic effects, which entail the release of ordered water molecules from the inner cavity of the cucurbituril host and de-solvation of the neutral guest molecules. Cationic guests can also be trapped inside cucurbiturils through cation–dipole and CT interactions.

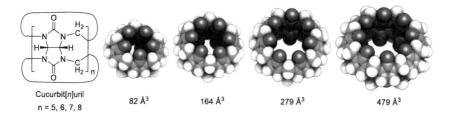

Cucurbit[n]uril
n = 5, 6, 7, 8   82 Å³   164 Å³   279 Å³   479 Å³

Figure 1.32 Molecular structures and CPK models and effective volumes of cucurbit[n] urils (n = 5–8).

1.5 Solvent Effects

Many chemical reactions and supramolecular assembling processes take place in a homogenous solution phase. The liquid component of a solution in the largest amount is called the **solvent**, while other substances in the solution are called **solutes**. Water is a "universal solvent" owing to its extremely large polarity ($\varepsilon = 80.1$ at 20 °C) and unique hydrogen bonding ability. In many chemical reactions, water is used to dissolve metal ions and polar compounds. In the biological world, water plays a life-sustaining role, supporting cellular structures and dictating the folding properties of amino acids, proteins, and DNA molecules to enable them to carry out their essential biological functions. Indeed, the impact of water as a solvent is so profound that we would generally classify solvents as **aqueous** or **nonaqueous**.

In the realm of organic chemistry, numerous nonaqueous solvents are used, since a large portion of organic compounds are hydrophobic and water insoluble [41]. Solvents can be classified by their physical and chemical properties. In general, liquids fall into three different classes, (1) molecular liquids, (2) ionic liquids, and (3) metals. Most of the commonly used liquid solvents used in chemical and biological applications are molecular liquids, which can be further classified as shown in Table 1.2.

Solvents play a vital role in organic reactions. As mentioned above, solvent molecules can interact with solute molecules in markedly different ways. The solvation effect of inert solvents (e.g., nonpolar alkanes) results in only minor changes in the structural and electronic properties of the solute molecules. Solvents with strong polarity and hydrogen bonding capability not only modify the properties of

Table 1.2 Classification of molecular solvents.

Classes	Properties	Examples
Inert	Unreactive, nonpolar, or weakly polar	Alkanes, fluorocarbons
Inert-polarizable	Electron pair donors, electron pair acceptors	Aromatic compounds, halogenated solvents, carbon disulfide, tetracyanoethane
Protogenic	Hydrogen bond donors	Acetic acid, sulfuric acid
Protophilic	Hydrogen bond acceptors	Tertiary amines
Amphiprotic	Having both hydrogen bond donor and acceptor capacity	Water, alcohols
Dipolar-aprotic	Have substantial dipole moments, but lack of marked hydrogen bond donor/acceptor capacity	Acetonitrile, DMF, DMSO, HMPA

solute molecules to a significant extent, but in some cases react with the solutes after dissolution. The solvent–solute reaction (i.e., solvolysis) changes the identity of the solute. Depending on the type of solvolysis reaction, a solvent molecule may act as a nucleophile or an electrophile and hence can strongly influence the kinetics and outcomes of the reaction. Take the S_N1 type of solvolysis of 2-admantyl triflate in water–ethanol mixtures as an example, the rate of solvolysis changes dramatically with the polarity of the mixture solvent (Figure 1.33A). An increase in the fraction of water from 20% to 80% leads to an acceleration of the reaction rate by three orders of magnitude. This dramatic solvent effect can be explained by the scheme illustrated in Figure 1.33B. The rate-limiting step of the S_N1 solvolysis is C–O bond breaking, yielding 2-adamantyl cation as the intermediate (IM). The transition state (TS) of this step bears a greater resemblance to the carbocation IM rather than the reactant, 2-adamantyl triflate. As discussed before, a cation–dipole interaction is stronger than a dipole–dipole interaction. It is therefore reasonable to deduce that in a polar medium, the solvation stabilization of the partially cationic TS is more pronounced than for the neutral reactant. As such, the energy barrier is lowered as the solvent polarity increases resulting in an increased solvolysis rate.

Solvolysis of various alkyl halides has been widely studied to evaluate the ability of the solvent to stabilize carbocations generated from a unimolecular rate-limiting step. Solvent effects also strongly influence other types of organic reactions. For example, if an S_N2 reaction uses an anion as the nucleophile, the reaction rate would be very slow in a protic solvent such as water or an alcohol, since the protic solvent molecules can act as hydrogen bond donors to strongly solvate the anion through hydrogen bonding interactions, which in turn reduce the nucleophilicity of the anion. The use of an aprotic dipolar solvent is preferred for such S_N2 reactions since it can sufficiently stabilize the transition state without compromising the reactivity of the anionic nucleophile. Solvents can also exert an

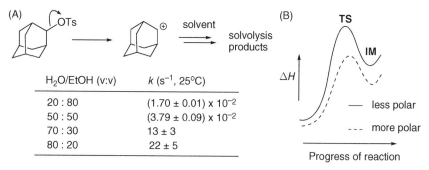

Figure 1.33 (A) Solvolysis of 2-admantyl triflate in mixtures of water and ethanol. Kinetic data are taken from *J. Org. Chem.* 1985, *50*, 4821–4823. (B) Schematic illustration of the solvent effects on the potential energy surface of the reaction.

important impact on acid–base exchange reactions. For example, the strength of a strong acid is limited or "leveled" by the basicity of the solvent. Similarly, the strength of a strong base is leveled by the acidity of the solvent. This is known as the **leveling effect**.

Solvent effects significantly influence CT interactions and related spectroscopic properties. As discussed before, the interactions of a donor (D) and an acceptor (A) molecule would lead to a CT complex, featuring a $D^{\delta+}\cdots A^{\delta-}$ type of charge separation. The stability of such a CT complex is solvent dependent. Polar solvents provide better stability for the CT complex than nonpolar solvents. The formation and properties of a CT complex can be probed by UV-Vis absorption analysis. In the UV-Vis absorption spectrum of a D⋯A complex, a new absorption band known as the **CT band** would emerge at a longer-wavelength position relative to the absorption bands of individual D and A molecules. The shape, intensity, and positions of UV-Vis absorption or fluorescence emission bands change as the nature of solvent is varied. This behavior is called **solvatochromism**. Monitoring the changes of the CT band of a chromophore in various solvents provides a quantitative way to assess solvent nature [42]. For example, the solvatochromic shifts of N-ethyl-4-methylcarboxypyridinium iodide (Figure 1.34A) in different solvents give the Z-scale for measuring solvent polarity. Herein Z is calculated by, $Z = 2.859 \times 10^4/\lambda_{max}$, where λ_{max} is the maximum absorption wavelength of the pyridinium iodide. Another popular set of solvatochromic parameters is called $E_T(30)$, which was first introduced by Reichardt et al. $E_T(30)$ is based on the electronic transition energy of a pyridinium-N-phenoxide dye (Figure 1.34B). The number 30 refers to the 30th candidate among numerous chromophores examined in their studies. The $E_T(30)$ scale can be expressed by the equation, $E_T(30) = 31.2 + 11.5\pi^* + 15.2\alpha$, where π^* reflects the polarizability and α describes the hydrogen bond-donating ability of the solvent.

Solvent effects are indispensable factors in supramolecular chemistry, especially in the formation of organized supramolecular self-assemblies through solvent-sensitive non-covalent forces such as hydrogen bonding, dipole–dipole interactions, and π–π stacking. Figure 1.35 shows an example of the solvophobically

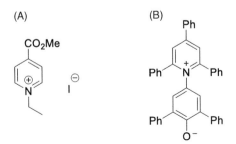

Figure 1.34 Two CT chromophores useful for evaluating solvent polarity.

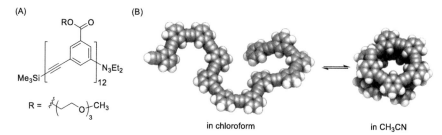

Figure 1.35 (A) Molecular structure and (B) the solvophobically driven conformational switching properties of an oligo(*m*-phenylene ethylene). Side chains and end groups are omitted in the illustration of molecular model for clarity.

driven folding of an oligomer of *m*-phenylene ethylene, which was synthesized and investigated by Moore and co-workers [43]. When dissolved in an appropriate solvent (chloroform), the oligomer prefers a random, disordered conformation. Upon switching to a more polar solvent (acetonitrile), the oligomer folds into a helical conformation, creating a cylindrical inner cavity that can be useful for hosting smaller guest species.

Further Reading

- Carey, F. A.; Sundberg, R. J., Advanced Organic Chemistry: Part A: Structure and Mechanisms. Springer Science & Business Media: 2007.
- Smith, M. B., March's Advanced Organic Chemistry: Reactions, Mechanisms, and Structure. John Wiley & Sons: 2020.
- Anslyn, E. V.; Dougherty, D. A., Modern Physical Organic Chemistry. University science books: 2006.
- Ozin, G. A.; Arsenault, A., Nanochemistry: A Chemical Approach to Nanomaterials. Royal Society of Chemistry: 2015.

References

1 Feynmann, R. P.; Leighton, R. B.; Sands, M., *The Feynmann Lectures on Physics*. Addison-Wesley: Reading, **1963**.
2 Taniguchi, N., On the Basic Concept of Nanotechnology. *Proc. of the ICPE*, Tokyo, **1974**, 18–23 .
3 Drexler, K. E., *Engines of Creation: The Coming Era of Nanotechnology*. Anchor Books: **1986**.
4 Lewis, G. N., The Atom and the Molecule. *J. Am. Chem. Soc.* **1916**, *38*, 762–785.
5 Gillespie, R. J.; Hargittai, I., *The VSEPR Model of Molecular Geometry*. Courier Corporation: **2013**.

6 Bader, R. F.; Gillespie, R. J.; MacDougall, P. J., A Physical Basis for the VSEPR Model of Molecular Geometry. *J. Am. Chem. Soc.* **1988**, *110*, 7329–7336.

7 Gillespie, R. J.; Robinson, E. A., Models of Molecular Geometry. *Chem. Soc. Rev.* **2005**, *34*, 396–407.

8 Cooper, D., *Valence Bond Theory*. Elsevier: Amsterdam, **2002**.

9 Pauling, L., The Nature of the Chemical Bond. Application of Results Obtained from the Quantum Mechanics and from a Theory of Paramagnetic Susceptibility to the Structure of Molecules. *J. Am. Chem. Soc.* **1931**, *53*, 1367–1400.

10 Pauling, L. C., The Theory of Resonance in Chemistry. *Proc. R. Soc.: Math. Phys. Sci.* **1977**, *356*, 433–441.

11 Gleiter, R.; Haberhauer, G., *Aromaticity and Other Conjugation Effects*. Wiley-VCH: Weiheim, **2012**.

12 Yates, K., *Hückel Molecular Orbital Theory*. Academic Press: New York, **1978**.

13 Krygowski, T. M.; Cyrański, M. K., Structural Aspects of Aromaticity. *Chem. Rev.* **2001**, *101*, 1385–1420.

14 Ho, T.-L., Hard Soft Acids Bases (HSAB) Principle and Organic Chemistry. *Chem. Rev.* **1975**, *75*, 1–20.

15 Pearson, R. G., Hard and Soft Acids and Bases. *J. Am. Chem. Soc.* **1963**, *85*, 3533–3539.

16 Mitchell, R. H., Measuring Aromaticity by NMR. *Chem. Rev.* **2001**, *101*, 1301–1316.

17 Chen, Z.; Wannere, C. S.; Corminboeuf, C.; Puchta, R.; Schleyer, P. V. R., Nucleus-Independent Chemical Shifts (NICS) as an Aromaticity Criterion. *Chem. Rev.* **2005**, *105*, 3842–3888.

18 Williams, R. V., Homoaromaticity. *Chem. Rev.* **2001**, *101*, 1185–1204.

19 Stahl, F.; Schleyer, P. v. R.; Jiao, H.; Schaefer, H. F.; Chen, K.-H.; Allinger, N. L., Resurrection of Neutral Tris-homoaromaticity. *J. Org. Chem.* **2002**, *67*, 6599–6611.

20 Rzepa, H. S., Möbius Aromaticity and Delocalization. *Chem. Rev.* **2005**, *105*, 3697–3715.

21 Ajami, D.; Oeckler, O.; Simon, A.; Herges, R., Synthesis of a Möbius Aromatic Hydrocarbon. *Nature* **2003**, *426*, 819–821.

22 Kawase, T.; Oda, M., Möbius Aromatic Hydrocarbons: Challenges for Theory and Synthesis. *Angew. Chem. Int. Ed.* **2004**, *43*, 4396–4398.

23 Yoon, Z. S.; Osuka, A.; Kim, D., Möbius Aromaticity and Antiaromaticity in Expanded Porphyrins. *Nat. Chem.* **2009**, *1*, 113--122.

24 Mulliken, R. S.; Rieke, C. A.; Brown, W. G., Hyperconjugation. *J. Am. Chem. Soc.* **1941**, *63*, 41–56.

25 Fleming, I., *Molecular Orbitals and Organic Chemical Reactions*. John Wiley & Sons: **2011**.

26 Alabugin, I. V.; Gilmore, K. M.; Peterson, P. W., Hyperconjugation. *Wiley Interdiscip. Rev. Comput. Mol. Sci.* **2011**, *1*, 109–141.

27 Alabugin, I. V.; Kuhn, L.; Krivoshchapov, N. V.; Mehaffy, P.; Medvedev, M. G., Anomeric Effect, Hyperconjugation and Electrostatics: Lessons from Complexity in a Classic Stereoelectronic Phenomenon. *Chem. Soc. Rev.* **2021**, *50*, 10212–10252.

28 von Ragué Schleyer, P.; Kos, A. J., The Importance of Negative (Anionic) Hyperconjugation. *Tetrahedron* **1983**, *39*, 1141–1150.

29 Garratt, P. J.; Ibbett, A. J.; Ladbury, J. E.; O'Brien, R.; Hursthouse, M. B.; Malik, K. A., Molecular Design Using Electrostatic Interactions. 1. Synthesis and Properties of Flexible Tripodand Tri-and Hexa-cations with Restricted Conformations. Molecular Selection of Ferricyanide from Ferrocyanide. *Tetrahedron* **1998**, *54*, 949–968.

30 Jeffrey, G. A., *An Introduction to Hydrogen Bonding.* Oxford University Press: New York, **1997**.

31 Bader, R. F., Atoms in Molecules. *Acc. Chem. Res.* **1985**, *18*, 9–15.

32 Arunan, E.; Desiraju, G. R.; Klein, R. A.; Sadlej, J.; Scheiner, S.; Alkorta, I.; Clary, D. C.; Crabtree, R. H.; Dannenberg, J. J.; Hobza, P., Definition of the Hydrogen Bond (IUPAC Recommendations 2011). *Pure Appl. Chem.* **2011**, *83*, 1637–1641.

33 Martinez, C. R.; Iverson, B. L., Rethinking the Term "Pi-stacking." *Chem. Sci.* **2012**, *3*, 2191–2201.

34 Hunter, C. A.; Sanders, J. K., The Nature of π-π Interactions. *J. Am. Chem. Soc.* **1990**, *112*, 5525–5534.

35 Pérez, E. M.; Martín, N., π–π Interactions in Carbon Nanostructures. *Chem. Soc. Rev.* **2015**, *44*, 6425–6433.

36 Soteras, I.; Orozco, M.; Luque, F. J., Induction Effects in Metal Cation–Benzene Complexes. *Phys. Chem. Chem. Phys.* **2008**, *10*, 2616–2624.

37 Stone, A., *The Theory of Intermolecular Forces.* 2nd ed.; Oxford University Press: Oxford, UK, **2013**.

38 Das, A.; Ghosh, S., Supramolecular Assemblies by Charge-Transfer Interactions between Donor and Acceptor Chromophores. *Angew. Chem. Int. Ed.* **2014**, *53*, 2038–2054.

39 Jiang, H.; Yang, X.; Cui, Z.; Liu, Y.; Li, H.; Hu, W.; Kloc, C., Adjusting Tetrathiafulvalene (TTF) Functionality through Molecular Design for Organic Field-Effect Transistors. *Cryst. Eng. Comm.* **2014**, *16*, 5968–5983.

40 Isaacs, L., Cucurbit[n]urils: From Mechanism to Structure and Function. *Chem. Commun.* **2009**, 619–629.

41 Reichardt, C.; Welton, T., *Solvents and Solvent Effects in Organic Chemistry.* John Wiley & Sons: Weinheim, **2011**.

42 Reichardt, C., Solvatochromic Dyes as Solvent Polarity Indicators. *Chem. Rev.* **1994**, *94*, 2319–2358.

43 Nelson, J. C.; Saven, J. G.; Moore, J. S.; Wolynes, P. G., Solvophobically Driven Folding of Nonbiological Oligomers. *Science* **1997**, *277*, 1793–1796.

2

Synthetic Methodologies for Preparation of Organic Nanomaterials

2.1 Organic Synthesis for Nanotechnology

In the preparation of nanoscale functional materials, not only precise control over size, dimension, and morphology must be carefully designed and implemented, but functionality, chemical reactivity, and stability at the molecular level should be taken into serious consideration. Practically, nanofabrication can be done through two approaches, namely top-down and bottom-up. The top-down approach does not have to deal with the molecular and/or atomic level of matter, but tends to encounter more and more difficulties when the size is continuously scaled down to the nanometer regime. The bottom-up approach in theory can circumvent this technical bottleneck but still face other challenges at both the molecular and supramolecular levels. The bottom-up approach starts from the use of small molecules as the basic building blocks, while the assembly of nanoscale objects via the bottom-up approach by a nanotechnologist is very little different from the undertakings of an organic synthetic chemist carrying out the synthesis of complex natural products. The toolbox for constructing complex organic nanomaterials, usually polymers and macromolecules, includes highly efficient and functional group tolerant reactions which can be repetitively executed to build up relatively large and complex molecular systems in high yields and without encountering significant difficulties in any of synthetic and purification steps. Beyond the scope of molecular synthesis, sophisticated nanoscale systems can also be assembled from certain molecular building blocks via specific non-covalent forces (hydrogen bonding, π–π stacking, etc.) commonly known as **self-assembly**. Over the past few decades, a vast array of synthetic methodologies has been well established by the synthetic community, empowering materials scientists and engineers to gain access to organic functional nanomaterials through reliable, high-yielding, and even more environmentally friendly reactions. This chapter will cover a collection of important covalent synthetic strategies and methodologies that have been very useful in the development of organic nanomaterials and molecule-based devices.

2.2 Planning the Synthesis

Chemistry is a creative science, in which the synthesis of molecules constitutes a dominant part. Chemical synthesis is a process of constructing complex molecular structures from small molecular building blocks, which are commercially available or easily accessible from natural resources. In essence, chemical synthesis is a bottom-up approach of building molecular entities. In the practice of organic synthesis, a **target molecule** (final product) is first decided based on the needs of industry (e.g., pharmaceuticals, dyes, polymers) or inspiration from naturally existing complex molecules (e.g., natural products). A detailed synthetic plan is subsequently developed and implemented with the aim of synthesizing the target molecule. Synthetic planning usually begins with the target molecule (i.e., final product) and then works backward to the starting materials using various reactions. Such a method for planning synthesis was first formulated by the famous American Chemist, Elias James Cory, who was awarded the 1990 Nobel Prize in Chemistry. Corey called his method **retrosynthesis** or **retrosynthetic analysis**, a method that can greatly simplify the synthetic planning for large complex molecules. Retrosynthetic analysis is based on known reactions. The synthetic plan generated from a retrosynthetic approach is used as a roadmap to guide the experimental synthetic work.

The basic concepts and terms of retrosynthesis are illustrated in Figure 2.1. In this example, a β-hydroxyketone is targeted for synthesis (called target molecule or TM). The retrosynthetic analysis begins with proposing the cleavage of a C–C bond β to the C=O group, referred to as a bond disconnection (BD) step. It is worth noting that BD is an imaginary bond breaking reaction that corresponds to the reverse of a real bond formation reaction. The BD step breaks down the TM into two smaller parts, which are hypothetical reagents called **synthons**. The synthons produced

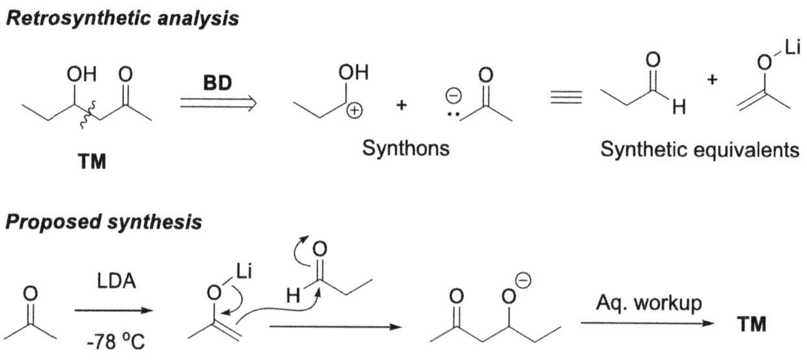

Figure 2.1 An example of retrosynthetic analysis.

in this retrosynthesis are a carbocation and an enolate ion, respectively. They are unstable and therefore cannot be used directly in a synthesis. However, they can be further converted into their synthetic equivalents, which are real reagents that carry out the same functions. In the retrosynthetic scheme, an open-ended retrosynthetic arrow (⇒) is used to indicate that the TM can be synthesized from the two synthons. Following the reverse direction of the retrosynthetic analysis, a plausible synthesis can be proposed.

In addition to bond disconnection (BD), other operations can be proposed in a retrosynthetic analysis such as functional group interconversion (FGI), functional group addition (FGA), and functional group removal (FGR). A proposal based on these operations must be consistent with known organic reactions, such as oxidation, reduction, substitution, addition, and elimination reactions. Therefore, it is important for a researcher to establish a sufficient knowledge base of synthetic methodologies before undertaking serious organic synthetic work.

For a complex TM, numerous plausible retrosynthetic analyses could be conceived and proposed. In planning a multistep synthesis, the overall yield of the target compound is dependent not only on the yield of each individual synthetic step, but the type of synthetic pathway adopted. In general, multistep synthesis can be carried out through three different types of pathways, namely linear synthesis, convergent synthesis, and divergent synthesis (outlined in Figure 2.2).

A **linear synthesis** follows a single pathway, in which all the intermediates and the TM are prepared in a sequential fashion. Although a linear synthesis is straightforward, it often suffers from a low overall yield if the synthetic pathway involves more than a few steps. Taking a five-step linear synthesis as an example, if each individual step affords a 90% yield, the overall yield is calculated as $(0.9 \times 0.9 \times 0.9 \times 0.9 \times 0.9) = 59\%$. However, if the yield of each step is dropped to 60%, the overall yield is then substantially reduced to $(0.6 \times 0.6 \times 0.6 \times 0.6 \times 0.6) = 7.7\%$. Comparatively, the convergent synthesis is more advantageous. In a **convergent synthesis**, the synthesis is broken down into two or more than two separate pathways. Each of the pathways involves a small number of individual steps. The products of these pathways are eventually joined together at the final stage of the synthesis to give the TM. As such, the convergent synthesis uses less materials and affords a relatively high overall yield. For example, the convergent pathway

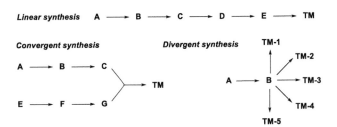

Figure 2.2 Illustrations of linear, convergent, and divergent synthesis.

shown in Figure 2.2 would give an overall yield of 73%, if each individual step gives a yield of 90%. If the yield of each individual step is 60% the overall yield is 34%, which is still much higher than that of the linear synthesis.

When the synthetic target is a family of compounds rather than a single structure, a divergent synthetic strategy (or **divergent synthesis**) is more advantageous than a linear synthesis. As shown in Figure 2.2, a divergent synthesis first produces a common intermediate, from which a multitude of products can be synthesized. Divergent synthetic methodology is particularly suitable for the studies aiming at the elucidation of structure–property relationships for a series of analogous molecular or macromolecular materials.

2.3 Useful Synthetic Methodologies for Organic Nanomaterials

2.3.1 Carbon–Carbon Bond Formation via Pd-Catalyzed Cross-Coupling Reactions

The formation of covalent bonds, particularly carbon–carbon bonds, is of the foremost importance in organic synthesis. There has been a vast body of synthetic methods documented in the synthetic literature to address the need for generating various types of carbon–carbon bonds. In classical organic synthesis, carbon nucleophiles (e.g., cyanide ion, Grignard reagents, organolithium reagents, and enolate ions) are used to react with carbon electrophiles (e.g., alkyl halides, carbonyl compounds, carboxylic acid derivatives, electron-deficient alkenes/arenes) for the formation of carbon–carbon bonds, which not only expand the size of the organic structures but introduce new functionalities and molecular properties to the products. In the field of organic materials research, the synthetic methods must be efficient, modular, and show broad applicability in many areas of research. In this light, catalytic reactions have become predominantly used for various synthetic undertakings, especially the preparation of π-conjugated organic materials. Many of these reactions occur through the coupling of two partners using transition metal catalysts, called **transition metal-catalyzed cross-coupling reactions** [1]. A general reaction scheme for cross-coupling reactions is shown in Figure 2.3, where the leaving group of an electrophile R_1–X (R_1 = alkenes or arenes, X = halides or pseudohalides) is substituted with a carbon nucleophile

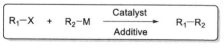

$R_1 = R_2$ = Alkenyl, aryl, etc.
X = Cl, Br, I, OTf, etc.
M = Mg, Zn, Zr, B, Cu, Sn, etc.
Catalyst = Pd, Ni, etc.

Figure 2.3 General reaction scheme for transition metal-catalyzed coupling reactions.

(R_2) typically in the form of an organometallic compound (M = Mg, Zn, Sn, etc.). The catalysts used for these couplings are usually Pd, Ni, and Cu-based complexes. The cross-coupling reaction results in a C–C bond linking the R_1 and R_2 units of the two coupling partners.

The original studies of transition metal-catalyzed cross-coupling reactions started in the late 1960s, when Pd was introduced as a catalyst. Ever since then, the field has undergone a rapid growth. Nowadays, numerous named cross-coupling reactions have been well established and widely adopted in academic research and industrial applications. Among them, the Mizoroki-Heck, Negishi, Suzuki (or Suzuki–Miyaura), Sonogashira, Stille, Hiyama, and Kumada-Tamao-Corriu couplings are the most distinguished examples. In 2010, the Nobel Prize in Chemistry was given to three synthetic chemists, Richard F. Heck, Ei-ichi Negishi, and Akira Suzuki, in recognition of their outstanding contributions to the development of Pd-catalyzed cross-coupling reactions.

Figure 2.4 lists the original examples of modern named cross-coupling reactions in a chronological order. As can be seen, these coupling reactions mainly differ in the types of R–M bonds in the nucleophilic coupling partners as well as the transition metals used for catalysis (Pd, Ni, etc.). The electrophilic partners are typically alkenyl and aryl halides/pseudohalides that contain a polar $C(sp^2)$–X bond, for which Pd-based catalysts show a high degree of effectiveness in promoting their cross-couplings. Ni-based catalysts also exhibit good catalytic activity for certain coupling reactions; however, they have not been as widely used as Pd-based catalysts. The R–M coupling partners (nucleophiles) can be diverse structures, in which the M unit is either a metal or a non-metal element that is more electropositive than carbon. For example, the M species used in the Kumada-Tamao-Corriu, Negishi, and Stille reactions are Mg, Zn, and Sn, respectively. Sonogashira cross-coupling uses an alkynyl copper(I) species *in situ* generated via the reaction of a terminal alkyne with a Cu(I) salt. In the cases of the Suzuki and Hiyama couplings, the nucleophile partners contain a carbon–boron bond and a carbon–silicon bond, respectively. For the cross-coupling reactions where acidic by-products (H–X) are formed, organic or inorganic bases need to be present. All such coupling reactions except the Heck reaction can be rationalized by a general simplified catalytic cycle as outlined in Figure 2.5. In this catalytic cycle, a Pd(0) species (L_nPd^0) acts as the active catalyst. In the first step, the organic halide/pseudohalide (e.g., Ar–X) reacts with the Pd(0) catalyst to form a Pd(II) complex, in which the Pd is inserted in between the former C–X bond. This step is therefore called **oxidative insertion**. In the second step, the Pd(II) complex exchanges the X group with the R group of the other partner through a step called **transmetallation** (i.e., metal exchange), resulting in another Pd(II) complex, in which the Ar and R groups are now both connected to the Pd(II) center. Finally, this complex undergoes a process called **reductive elimination** to afford the cross-coupled

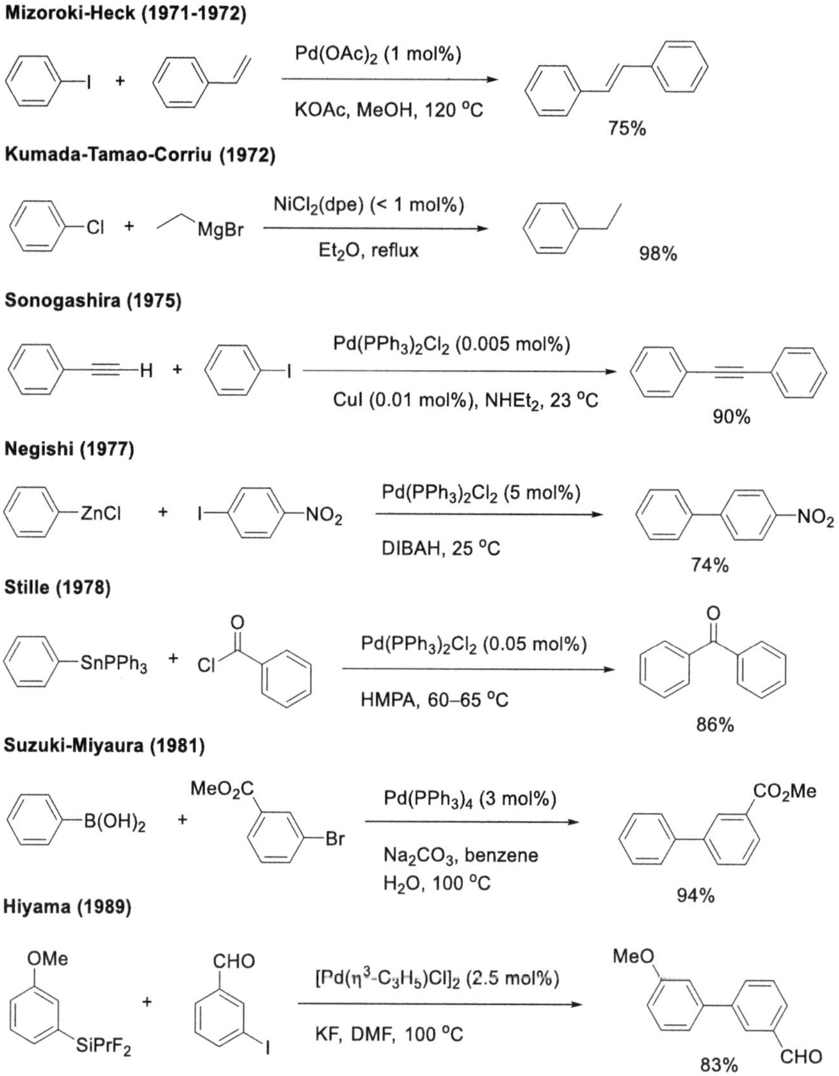

Mizoroki-Heck (1971-1972)

Kumada-Tamao-Corriu (1972)

Sonogashira (1975)

Negishi (1977)

Stille (1978)

Suzuki-Miyaura (1981)

Hiyama (1989)

Figure 2.4 Examples of named transition metal-catalyzed cross-coupling reactions.

product (Ar–R) and to regenerate the L_nPd^0 catalyst to be carried on to the next catalytic cycle.

Mizoroki-Heck cross-coupling (often referred to as the Heck reaction) uses an alkene and an alkenyl or aryl halides/pseudohalide as the coupling partners. Its mechanism differs from the other Pd-catalyzed cross-coupling reactions. As shown in Figure 2.6, after the oxidative insertion step, the L_nXPd^{II}–Ar complex reacts with the alkene substrate through a migratory insertion. The resulting

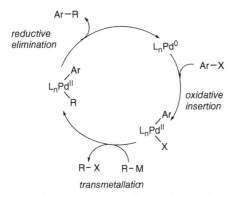

Figure 2.5 A generic catalytic cycle for Pd-catalyzed cross-coupling reactions.

Figure 2.6 The catalytic cycle for the Heck reaction.

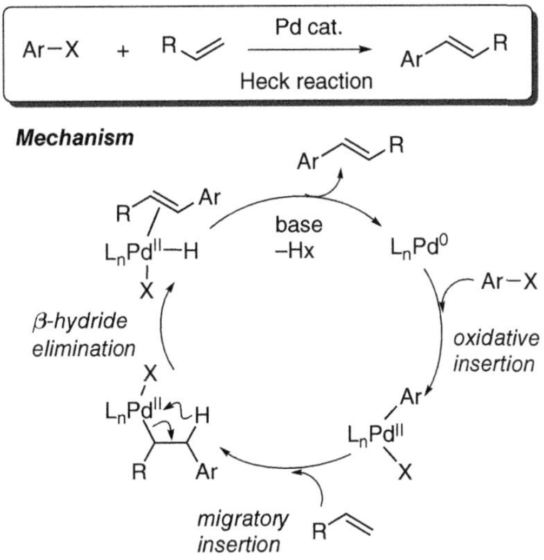

intermediate then undergoes by a *syn*-periplanar β-hydride elimination event to give a Pd/alkene complex, which eventually loses a molecule of H–X in the presence of a base to yield the cross-coupled product and to regenerate the active Pd(0) catalytic species. Because of the lack of a reductive transmetallation step in the catalytic cycle, it has been argued that the Heck reaction is not a true cross-coupling reaction.

Pd catalysts play an extremely important role in the cross-coupling reactions. One commonly used Pd(0) catalyst is Pd(PPh$_3$)$_4$, which can very efficiently catalyze cross-coupling reactions. However, this Pd complex is not very stable when exposed to air and moisture. Therefore, it needs to be stored under moisture and air-free conditions. To circumvent this problem, stable Pd(II) salts such as PdCl$_2$, Pd(PPh$_3$)$_2$Cl$_2$, and Pd(OAc)$_2$ are often utilized as pre-catalysts. They can be converted to active Pd(0) species *in situ* during cross-coupling reactions. Certain ligands can bring better stability to the Pd(0) complexes. For example, dibenzylideneacetone (dba) as the ligand gives an air-stable complex, Pd(dba)$_2$. Moreover, tuning of the steric and electronic properties of the ligand has been investigated as a major driving force for the improvement of catalytic activity, as ligands have been found to have a significant effect on the steps of the catalytic cycle. For instance, strong σ-donating ligands (e.g., PR$_3$, R = alkyl) increase electron density around the metal and hence can accelerate the oxidative insertion step, which is commonly believed to be the rate-determining step in many cross-coupling reactions. The rate of the elimination step is accelerated by the use of bulky ligands exhibiting a large cone angle (also known as the **Tolman angle**, Figure 2.7A). For example, Pd(P^{t}Bu$_3$)$_2$ and Pd[**P(o-tolyl)$_3$**]$_2$ are highly effective catalysts for a wide range of cross-coupling reactions. Bidentate ligands with wide **bite angles** have also been frequently utilized in Pd catalysts. This type of ligand exhibits inherently strong chelation effects and high stability, while the cone and bite angles of the catalyst (Figure 2.7A) can generate a significant impact on the catalytic process.

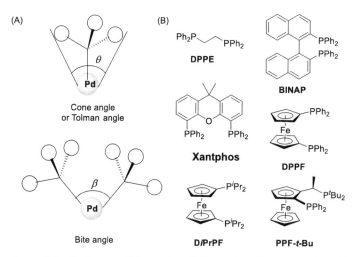

Figure 2.7 (A) Cone and bite angles. (B) Examples of bidentate phosphine ligands for Pd catalysts.

Pd-catalyzed cross-coupling reactions have been extensively applied to the synthesis of organic π-conjugated oligomers and polymers with intriguing nanoscale structures and electronic properties. Figure 2.8 gives three synthetic examples of π-conjugated co-polymers using named cross-coupling reactions. During polymerization, reactive aryl halides are cross-coupled with corresponding partners either present in the same monomer (Figure 2.8A) or in another monomer (Figure 2.8B and 2.8C). These polymerization reactions yield polydispersed products (i.e., mixtures of polymers with varied chain lengths).

To obtain π-conjugated oligomers with precisely controlled structures and monodispersity, an **iterative divergent/convergent** (IDC) strategy can be used [2]. Figure 2.9 shows an example of using Sonogashira cross-coupling in

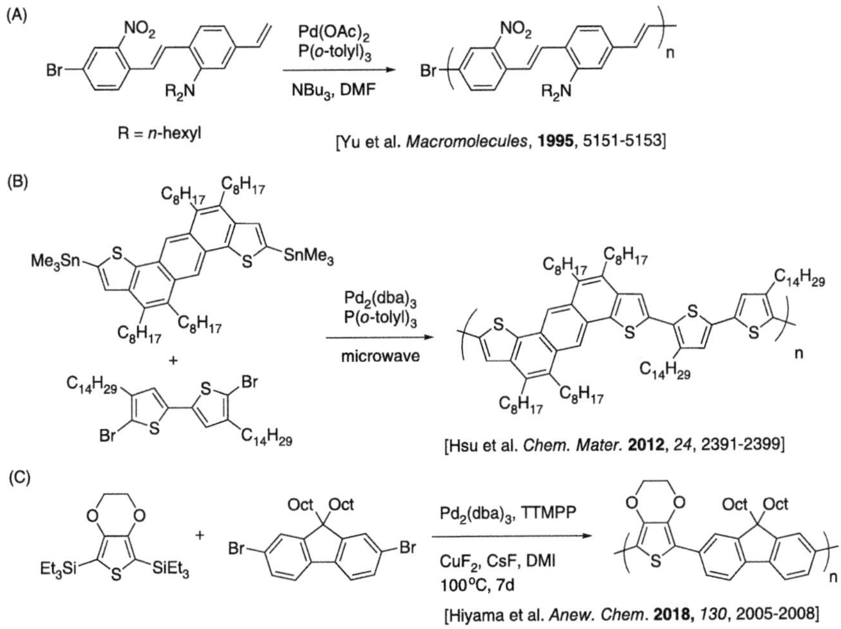

Figure 2.8 Synthesis of π-conjugated polymers using (A) Heck coupling, (B) Stille coupling, and (C) Hiyama coupling as the key steps.

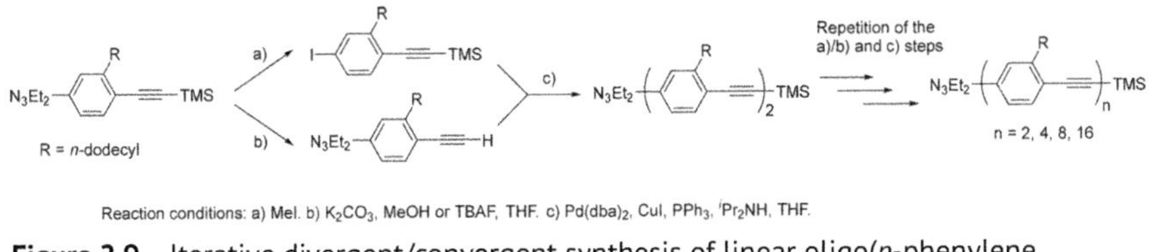

Reaction conditions: a) MeI. b) K₂CO₃, MeOH or TBAF, THF. c) Pd(dba)₂, CuI, PPh₃, ⁱPr₂NH, THF.

Figure 2.9 Iterative divergent/convergent synthesis of linear oligo(p-phenylene ethynylene)s using Sonogashira cross-coupling as the key C–C bond forming step.

combination with prudently designed protection/deprotection sequences to rapidly produce long monodispersed oligo(p-phenylene ethynylene)s through the IDC approach [3]. The synthesis begins with converting an orthogonally protected phenylene ethynylene monomer into two building blocks through two parallel steps (divergent synthesis). Step a) transforms the Et_2N_3 end group to an aryl iodide using MeI, while step b) is a protiodesilylation of a trimethylsilyl (TMS)-protected alkyne group by K_2CO_3 or n-tetrabutylammonium fluoride (TBAF). The two intermediates are then subjected to the Sonogashira coupling (convergent synthesis) catalyzed by $Pd(dba)_2$/CuI in the presence of PPh_3/$^iPr_2NH$ to yield a phenylene ethynylene dimer. Simply repeating the divergent and convergent steps leads to the formation of 4-mer, 8-mer, and 16-mer. It is well worth mentioning that the 16-mer has a molecular weight of 4,461 Da and its chain length reaches about 128 Å in its extended conformation, which is sufficiently long to bridge the lithographically derived probe gaps of nanoelectronic devices.

Cross-coupling reactions can even be performed at a liquid/solid interface to induce versatile functionalization of nanoparticles and precise patterning of self-assembly covered surfaces. Figure 2.10 illustrates an elegant synthesis that uses a Suzuki reaction to modify supramolecular self-assembled monolayers (SAMs) [4]. In this work, a gold surface was first coated with a monolayer of aryl bromide and then incubated in a methanolic solution of phenylboronic acid and sodium acetate at a millimolar concentration. An atomic force microscopic (AFM) scan was performed on the surface with a Pd nanoparticle-coated tip. The Pd nanoparticles acted as a catalyst to induce Suzuki coupling between aryl bromide and phenylboronic acid within the area where the AFM tip scanned. As such, cross-coupling reactions were performed in a spatially resolved manner on the SAM covered surface.

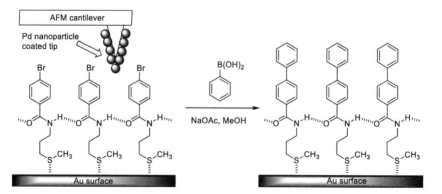

Figure 2.10 Cross-coupling of phenylboronic acid with the SAM of aryl bromides through the Suzuki reaction initiated by a Pd nanoparticle-coated AFM tip.

2.3.2 Carbon-Carbon Bond Formation via Other Types of Cross-Coupling Reactions

In addition to traditional cross-coupling reactions, recent research has disclosed many new classes of TM-catalyzed cross-coupling reactions. One type of cross-coupling reaction that has been extensively developed in recent years is called Pd-catalyzed oxidative cross-coupling, which is particularly useful for making arene–arene linkages. Unlike traditional cross-coupling reactions, this type of reaction does not need an aryl halide/pseudohalide (Ar–X) and an organometallic (R–M) partner. Instead, the cross-coupling takes place directly between two arenes through selective C–H activations. As shown in Figure 2.11, the catalytic cycle begins with two selective C–H insertion steps, leading to a diaryl-Pd(II) complex that easily undergoes reductive elimination. The reductive elimination produces the cross-coupled diaryl product and a Pd(0) species. Re-oxidation of the Pd(0) then completes the catalytic cycle. Pd-catalyzed oxidative cross-coupling reactions require experimental conditions to be carefully controlled to avoid the homocoupling reactions. To address this challenge, innovative solutions have

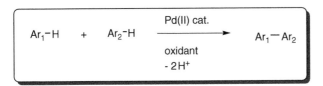

Mechanism

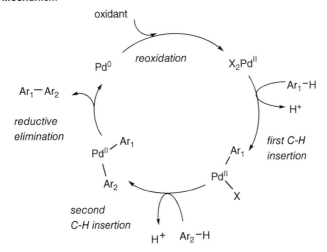

Figure 2.11 The general catalytic cycle of Pd(II)-catalyzed oxidative cross-coupling reactions.

been proposed and investigated. Cross-coupling reactions between two inactivated arenes can now be possibly achieved in good yields (see examples in Figure 2.12).

In contrast to the above listed reactions, TM-catalyzed cross-coupling reactions can proceed between two electrophilic partners through a so-called reductive cross-coupling pathway. This type of cross-coupling reaction can be promoted by a range of TM catalysts, including Co-, Ni-, Pd-, and Fe-based systems. Representative reactions are shown in Figure 2.13. As can be seen, the coupling reactions are effectively carried out using less expensive TM-catalysts (e.g., Ni, Co) in addition to Pd-catalysts. Also, the catalytic processes involve the use of a reducing agent (e.g., Mn, Zn) to re-activate the TM-catalyst at the end of the catalytic cycle. The reductive cross-coupling reactions also allow for the formation of $C(sp^2)$–$C(sp^3)$ bonds, giving access to a wide array of products. Finally, the reduction step can be designed to occur through photochemical and electrochemical means. This can provide exciting opportunities for exploring new chemistry and result in the discovery of useful synthetic methodologies.

Another class of cross-coupling reactions, namely **direct arylation** [5, 6], has also received growing attention in synthetic applications of C–C bond formation. Direct arylation in general involves the cross-coupling of an activated arene and an inactivated arene or heteroarene. It is therefore viewed as a type of C–H activation synthesis. As shown in Figure 2.14, the inactivated arene can carry a directing group (DG), typically an oxygen or nitrogen tether, to guide the transition metal into place. In the absence of a DG, regioselectivity can be controlled by the acidity of the C–H bond through the substituent effects (electron-donating or withdrawing) on the arene substrate. Electron-rich heteroarenes (e.g., S, O, N

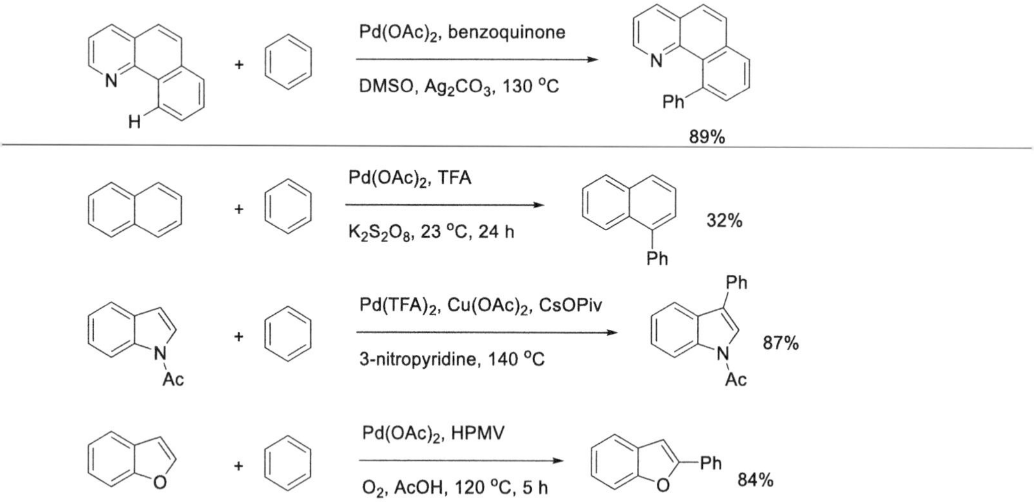

Figure 2.12 Selected examples of Pd(II)-catalyzed oxidative cross-coupling reactions.

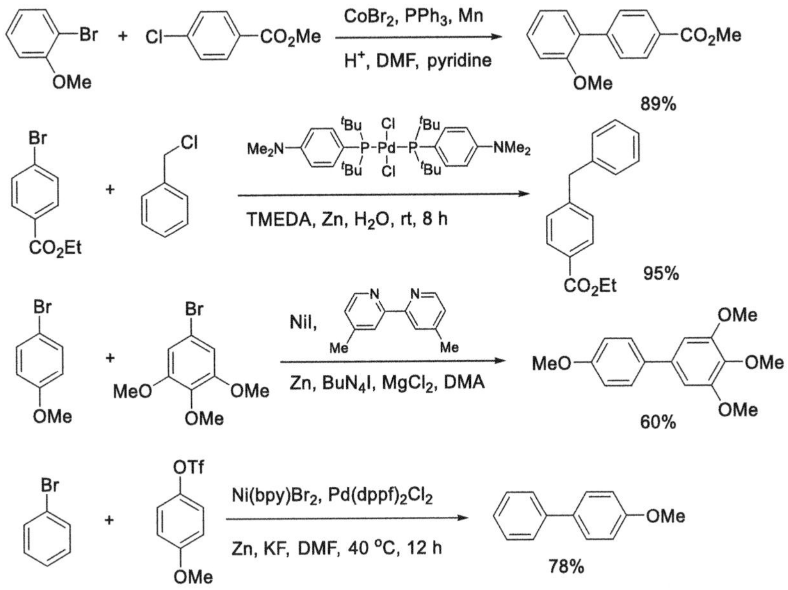

Figure 2.13 Selected examples of TM-catalyzed reductive cross-coupling reactions.

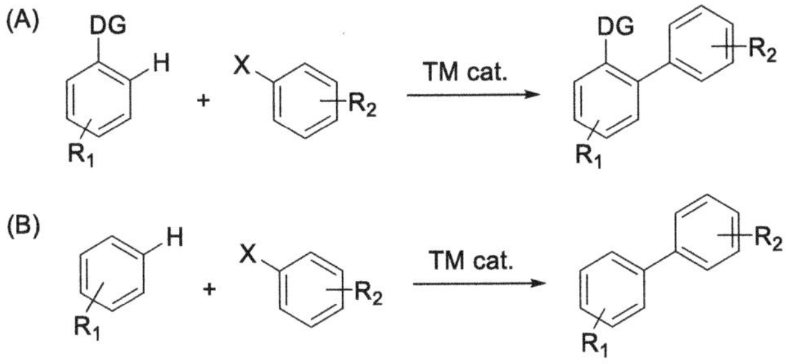

Figure 2.14 General types of intermolecular direct arylation.

containing heterocycles) can also be used for direct arylation. Many transition metal complexes effectively catalyze direct arylation processes, including Pd, Ru, and Rh catalysts. Typically, the reactions catalyzed by a Pd complex involve Pd(0) and Pd(II) species. In addition to intermolecular reactions, the direct arylation can also occur in an intramolecular fashion, facilitating the synthesis of a wide scope of polycyclic aromatic hydrocarbons (PAHs). Selected examples of direct arylation reactions are illustrated in Figure 2.15.

Compared with traditional TM-catalyzed cross-coupling reactions, the direct arylation approach has a number of advantages, such as being more atom-economical, requiring reduced synthetic effort in preparing the activated

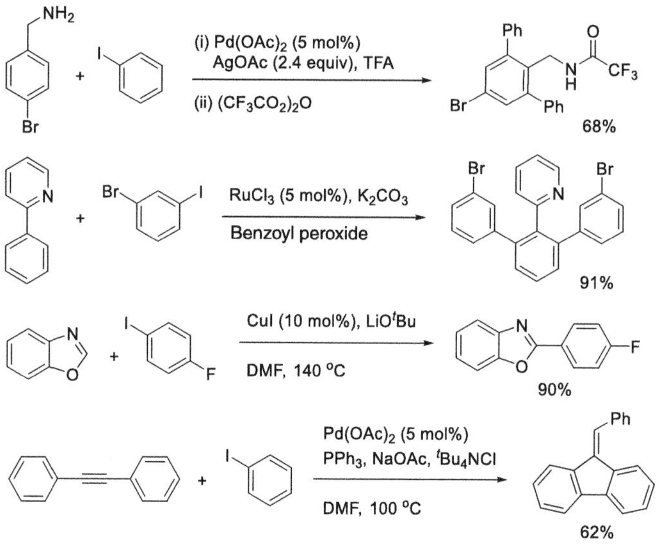

Figure 2.15 Examples of TM-catalyzed direct arylation reactions.

coupling partners, and having less problems associated with side reactions and/or by-product formation. For these reasons, direct arylation has been used as an efficient and environmentally friendly approach for the preparation of functional π-conjugated oligomers and polymers. For example, Liu and co-workers utilized a one-pot polymerization (Figure 2.16) to achieve rapidly access to a type of linear D–A π-conjugated oligomer with chain lengths ranging from 3 to 10 nm [7].

2.3.3 Carbon–Carbon Bond Formation through TM-Catalyzed Homocoupling Reactions

Homocoupled by-products are frequently observed in TM-catalyzed cross-coupling reactions. Even though optimizing reaction conditions to minimize the formation of homocoupled products is an important part of synthetic efforts in cross-coupling reactions, the application of TM-catalyzed homocoupling reactions as a C–C bond forming strategy has been shown very useful in the preparation of functional organic macromolecules and nanomaterials. Compared with cross-coupling reactions, the synthetic scope of homocoupling reactions is mainly limited to the preparation of structurally symmetrical target molecules. Nevertheless, some homocoupling reactions obtain starting materials with ease and are highly efficient when applied in special types of synthesis, such as macrocyclization, polymerization, and molecular recognition processes. For these reasons, TM-catalyzed homocoupling reactions have also been popularly adopted in organic nanochemistry.

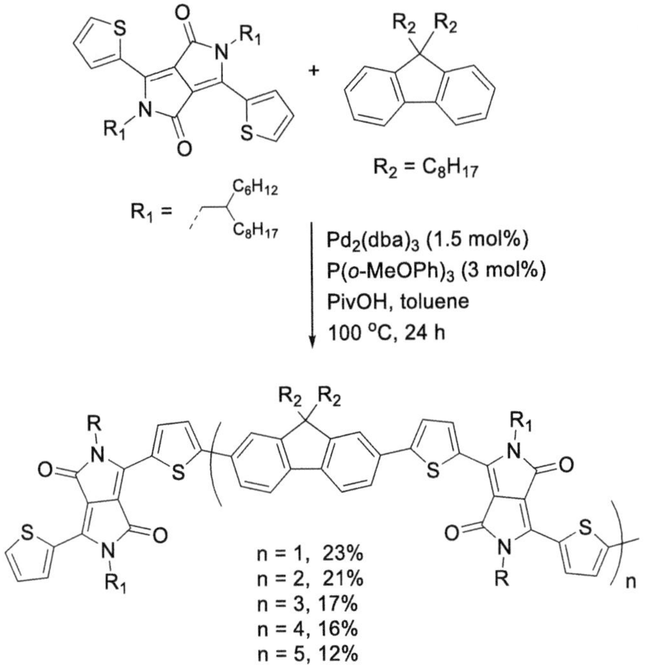

Figure 2.16 Rapid synthesis of nanoscale π-conjugated oligomers through Pd-catalyzed direct arylation reactions.

A classical homocoupling reaction is the well-known Glaser coupling. As shown in Figure 2.17, two molecules of a terminal alkyne can be coupled to form a 1,3-butadiyne product in the presence of a Cu(I) salt and an amine. Beside forming a C(sp)–C(sp) bond, the homocoupling reaction yields water as the only by-product. It has been shown that the presence of dioxygen (O_2) is necessary to this type of reaction. Cu(I) is predominantly used as the catalyst for this reaction, but it is relatively unstable and can be easily oxidized to Cu(II). Therefore, a reducing agent is usually added to bring the Cu(II) back to Cu(I). The Glaser reaction was first reported in 1869. Hay modified the reaction conditions in 1962 with the addition of a catalytic amount of N,N,N',N'-tetramethylethylenediamine (TMEDA) in the presence of O_2 [8]. This improved method has been often referred to as the Glaser-Hay coupling. Nowadays, the Glaser reaction has been further developed with numerous modifications, in which a variety of catalysts (e.g., Cu/CuO$_2$ nanoparticles, CuBr, CuI, and Cu(OAc)$_2$ are used in combination with carefully chosen ligands.

Figure 2.18A demonstrates the power of the Glaser reaction in constructing linearly π-conjugated oligoynes. In this reaction, a silyl-protected triyne precursor

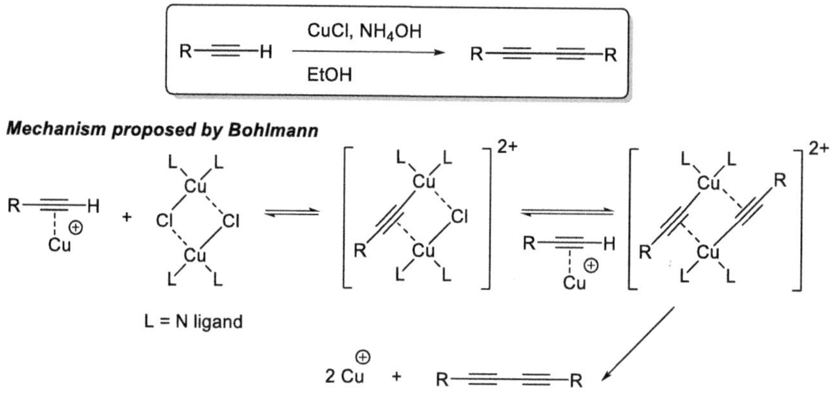

Figure 2.17 Glaser coupling and the mechanism proposed by Bohlmann [9].

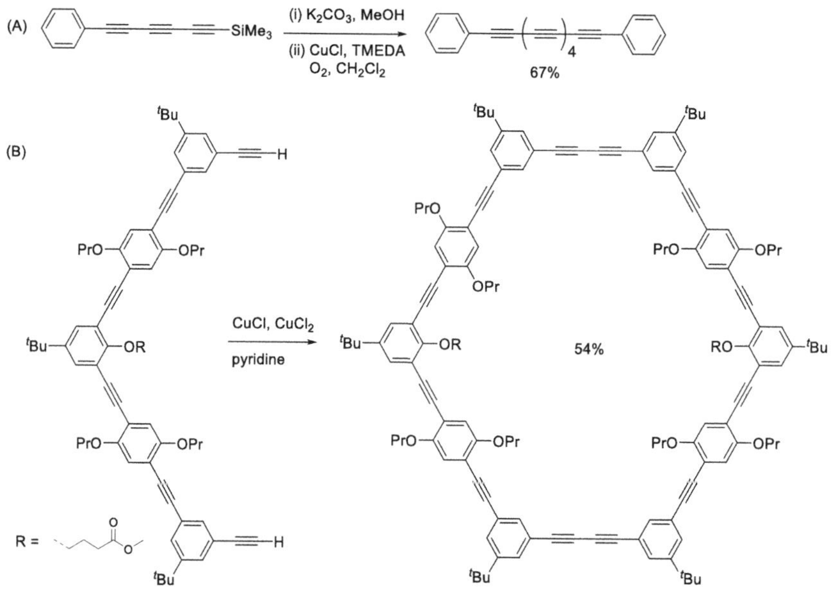

Figure 2.18 Synthesis of (A) a hexayne molecular wire and (B) a *m*-phenylene ethynylene shape-persistent macrocycle using the Glaser-Hay homocoupling as the key synthetic step.

first undergoes a protiodesilylation with K_2CO_3 in MeOH. Usually, the protiodesilylation proceeds in a nearly quantitative yield such that the resulting terminal alkyne intermediate can be directly subjected to the Glaser-Hay homocoupling reaction without the need for further purification. The reaction leads to the formation of a hexayne unit, which can be viewed as a linear carbon nanowire.

In another example (Figure 2.18B), a wedge-shaped *m*-phenylene ethynylene precursor is treated with $CuCl/CuCl_2$ in the presence of pyridine to form a cyclic product belonging to the family of **shape-persistent macrocycles** [10]. Shape-persistent macrocycles show a rigid cyclic skeleton made of a regular repeating unit with a low degree of conformational freedom. As such they can orderly stack to form defined channels to host small guest molecules. The homocoupling of terminal alkynyl functionalized precursors is a very attractive synthetic approach for shape-persistent macrocycles, since the linear C(sp)–C(sp) bond extension provides straightforward shape control and usually affords good yields for macrocyclization.

It is quite remarkable that Glaser coupling can even be applied to the synthesis of structurally complex assemblies; for instance, Tykiwinksi and Anderson utilized a phenanthroline macrocycle to explore the Cu-templated synthesis of a series of rotaxanes (Figure 2.19) [11].

The homocoupling of terminal alkynes can be efficiently promoted under catalytic conditions which are similar to those used in Sonogashira cross-coupling reactions. As a matter of fact, it has been known that some Sonogashira cross-coupling reactions are plagued by homocoupled by-products, particularly when Pd(II) is used as the pre-catalyst and air is not thoroughly removed from the reaction vessel. The key steps in Pd-catalyzed alkyne homocoupling reactions are the transmetallation of $Pd(II)X_2$ with two molecules of alkynyl cuprate to form a dialkynyl-Pd complex, which subsequently undergoes a reductive elimination to afford a homocoupled alkyne product and a Pd(0) species. It is essential for the reaction to have an oxidizing agent (e.g., air, oxygen, iodine, quinones) present, so that the Pd(0) can be transformed back into Pd(II) to complete the catalytic cycle. Pd-catalyzed alkyne homocoupling reactions have found extensive applications in making conjugated oligoyne/polyyne derivatives, shape-persistent macrocycles, and π-conjugated arylene butadiynylene oligomers and polymers. As shown in

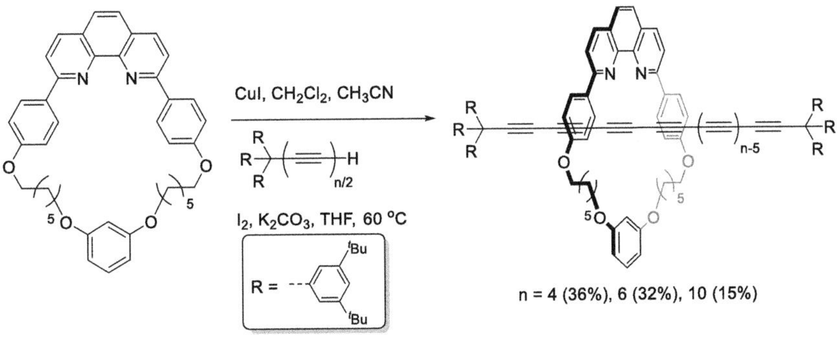

Figure 2.19 Cu-templated synthesis of rotaxanes consisting of a phenanthroline macrocycle threaded with various linear polyynes.

Figure 2.20A, alkyne homocoupling reactions can be performed immediately after a Sonogashira coupling in a one-pot fashion, eventually giving a butadiyne product in an excellent yield. Structurally rigid dialkynyl-arenes can undergo macrocyclization to form shape-persistent macrocycles with pre-designed shapes. For example, in Figure 2.20B, a group of diethynyl-pyrenes is subjected to Pd(II)-catalyzed homocoupling macrocyclization in the presence of Cu(I) as a co-catalyst and iodine as an oxidant [12]. The organization of the resulting macrocycle is predetermined by the alignment of the dialkynyl groups in the pyrene precursors. Polymerization reactions can also be carried out through the Pd(II)-catalyzed homocoupling. Figure 2.20C illustrates the synthesis of a class of conjugated co-polymers containing tetrathiafulvalene vinylogue (TTFV) and phenylacetylene building blocks in their molecular backbones [13]. Since the precursors are preorganized in a wedge-shaped conformation, their structural extension through alkynyl homocoupling eventually leads to a coiled polymeric framework. As such, the homocoupling polymers can be applied to efficiently wrap around individual single-walled carbon nanotubes (SWNTs) with certain diameter selectivity.

Besides alkyne homocoupling reactions, aryl electrophiles, catalyzed by TM catalysts, can form biaryl products through homocoupling reactions. The classical version of this type of reaction is known as the Ullmann coupling. As shown in Figure 2.21, various aryl halides can undergo homocoupling in the presence of

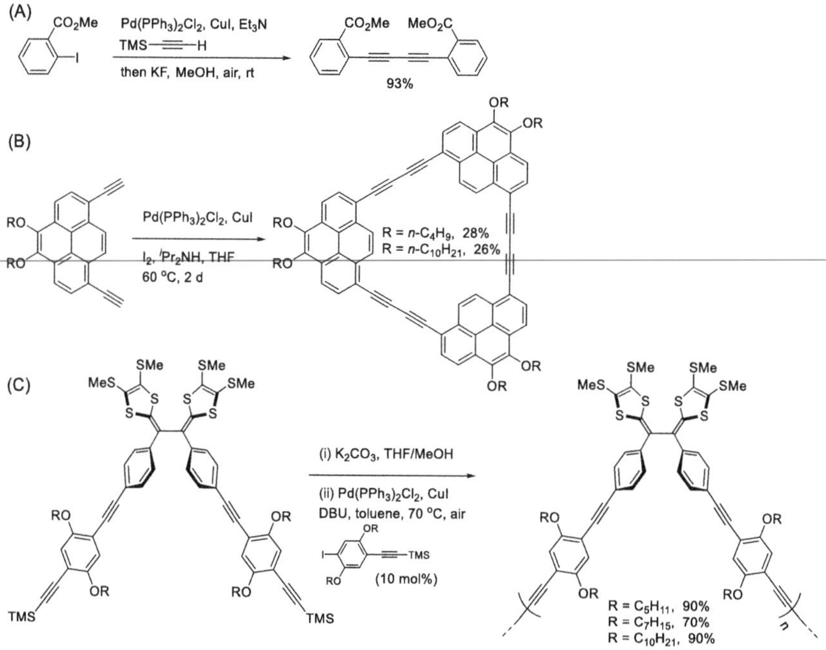

Figure 2.20 Examples of Pd-catalyzed alkyne homocoupling reactions.

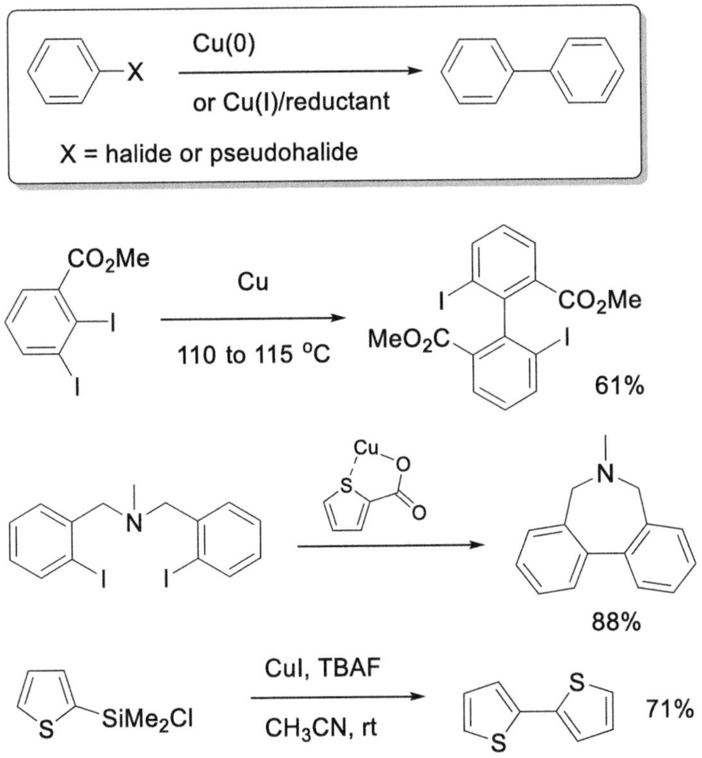

Figure 2.21 Examples of Cu-mediated Ullmann coupling reactions.

Cu(0) or a Cu(I) salt, usually present in a stoichiometric or an excess amount. The mechanism for Cu(0)-mediated Ullmann coupling involves the formation of aryl radicals, while the mechanism for Cu(I)-induced homogeneous Ullmann coupling follows a pathway involving a Cu(III) species. Traditional Ullmann coupling reactions use excess Cu(0) and high temperature. The reactions proceed most rapidly with aryl halides substituted at the *ortho* position. The use of homogeneous Cu(I) catalysts (e.g., Cu(OTf) and CuI) allows the coupling reactions to occur under much milder conditions. The introduction of Ni(0) catalysts in the early 1970s represented a significant advance in this type of reaction. Ni(0) catalysts can be obtained from Ni complexes, such as Ni(cod)$_2$ and Ni(PPh$_3$)$_4$, or from *in situ* reductions of some Ni(II) salts.

An elegant application of the Ni(0)-catalyzed Ullmann coupling in organic material synthesis is highlighted in Figure 2.22. Krische and co-workers in 2018 performed a threefold reductive biaryl homocoupling to construct a type of helical rodlike molecular cage in 17% yield [14]. Given that the formation of the product undergoes three steps of biaryl homocoupling with specific regioselectivity, this yield is considered remarkable.

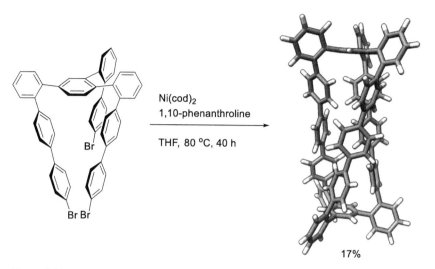

Figure 2.22 Synthesis of a helical rodlike phenylene cage through Ni(0)-catalyzed homocoupling.

2.3.4 Alkene and Alkyne Metathesis Reactions

Apart from C–C single bond forming processes, the generation of C=C and C≡C bonds is also vitally important in modern organic synthesis. Traditional synthetic methods utilize elimination and rearrangement reactions to produce C=C and C≡C bonds. Over the past decades, TM-catalyzed metathesis reactions have emerged and grown rapidly, offering powerful C=C and C≡C forming methods. The successful development of such synthetic methodologies was recognized by the 2005 Nobel Prize in Chemistry awarded jointly to Yves Chauvin, Robert H. Grubbs, and Richard R. Schrock for the development of the metathesis method in organic synthesis. Nowadays these reactions are widely used in various synthetic schemes [15–17].

Metathesis is a Greek word that means "transposition." Alkene metathesis, also called olefin metathesis, is the process of breaking the C=C bonds in two alkenes and then joining the fragments to generate new alkene products. Figure 2.23 shows a general alkene metathesis reaction and a well-established mechanism known as the Chauvin mechanism. Synthetically, terminal alkenes are favored as reactants, since their metathesis reactions give ethene as a by-product, which can be easily removed from the reaction mixture. As such, the equilibrium of the metathesis reaction strongly favors the formation of a new alkene product and ethene as by-product.

The olefin metathesis reactions fall into several important classes, including **cross metathesis** (CM), **ring-opening metathesis** (ROM), **ring-closing**

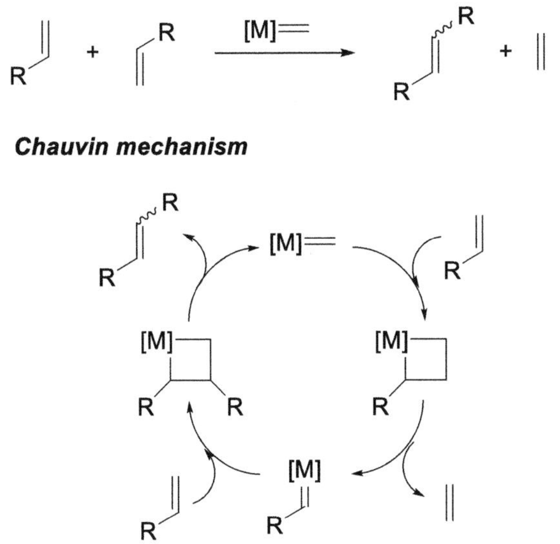

Chauvin mechanism

Figure 2.23 General TM-catalyzed reaction scheme for olefin metathesis.

metathesis (RCM), **ring-opening metathesis polymerization** (ROMP), and **acyclic diene metathesis** (ADMET). The most useful catalysts are the Schrock catalyst, the first and second-generation Grubbs catalysts, and the Blechert-Hoveyda catalyst (Figure 2.24). In recent years, a great number of new catalysts have also been developed, primarily aimed at increasing catalyst stability and efficiency. Most of the catalysts are Ru and Mo complexes, in which the ligands play important roles in dictating the catalytic performance. It is generally known that the Ru catalysts are compatible with alcohols, carboxylic acids, and aldehydes, but not with amines and phosphanes. On the other hand, the Mo catalysts are compatible with amines and phosphanes, but lack tolerance toward alcohols, carboxylic acids, and aldehydes. Therefore, a prudent choice of a catalyst must be made before planning an olefin metathesis reaction. The advances in this field have so profoundly impacted modern organic synthetic chemistry that many challenging syntheses, for example, tri- and tetrasubstituted olefins with well-defined stereochemistry, can now be readily addressed using this metathesis strategy.

In the field of organic functional materials, olefin metathesis reactions have been found to show a high degree of efficiency when used to synthesize challenging macromolecular structures. Figure 2.25A illustrates an efficient one-pot macrocyclization to generate an *m*-phenylene vinylene-based shape persistent macrocycle from a simple *m*-divinylbenzene precursor [18]. Recently, Wang and co-workers took advantage of the catalytic power of the second-generation Grubbs catalyst to prepare a class of carbon nanobelts [19]. As exemplified in Figure 2.25B,

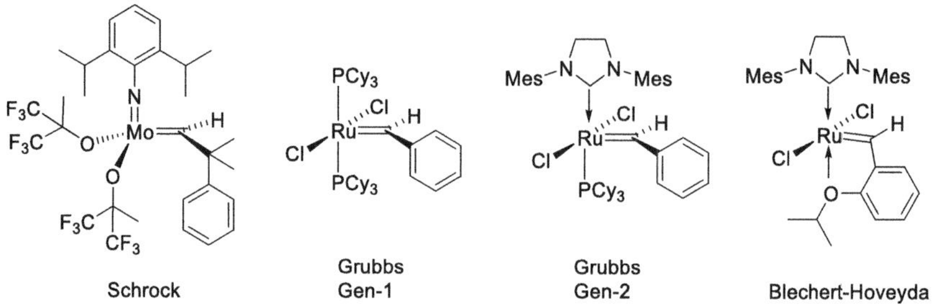

Figure 2.24 Classical olefin metathesis catalysts.

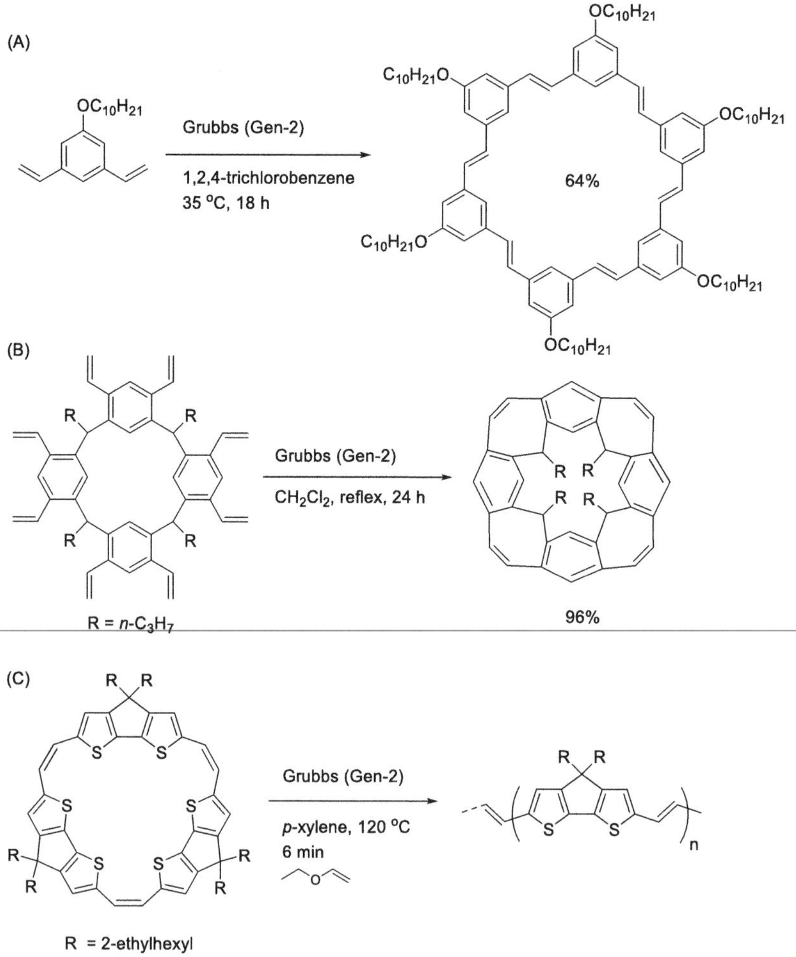

Figure 2.25 Examples of synthesis of organic π-conjugated molecules and polymers through olefin metathesis reactions.

the metathesis reactions result in ring closure to generate the rim of the nanobelt in a high yield. Ring-opening metathesis polymerization (ROMP) is a popular methodology used in the preparation of polymeric materials and structures. In the example shown in Figure 2.25C, linear π-conjugated polymers are produced through a ROMP approach using a cyclic triene as the precursor [20].

Like alkenes, alkynes can also undergo metathesis reactions catalyzed by certain transition metal complexes [21–23]. Figure 2.26 shows a general alkyne metathesis reaction and the mechanism associated with it. Compared with the olefin metathesis reactions, alkyne metathesis reactions are less studied. Active exploration of their scope and applications in synthesis is still ongoing. The original alkyne metathesis reaction was reported as early as 1974. The reaction occurred in homogeneous phase, using $Mo(CO)_6$ and a phenol additive at high temperature (160 °C), making its synthetic application quite limited [24]. The introduction of high-valent transition-metal alkylidyne complexes, known as Schrock alkylidynes, has greatly promoted the development of active catalysts for alkyne metathesis reactions. Today, the widely used catalysts are variants of the Schrock alkylidynes, which contain a M≡C motif, where the metal (M) can be tungsten, molybdenum, or ruthenium.

The mechanistic steps shown in Figure 2.26 are inherently reversible. Usually, methyl-capped alkynes are preferred substrates for the metathesis reactions, since the butyne produced, although not highly volatile, can be readily trapped in molecular sieves to drive the equilibrium in the desired direction. Additionally, the formation of insoluble alkyne products (precipitation) has been demonstrated effective at driving the alkyne metathesis reactions. Figure 2.27 shows the synthesis of a carbazole-ethynylene-containing shape-persistent macrocycle through two

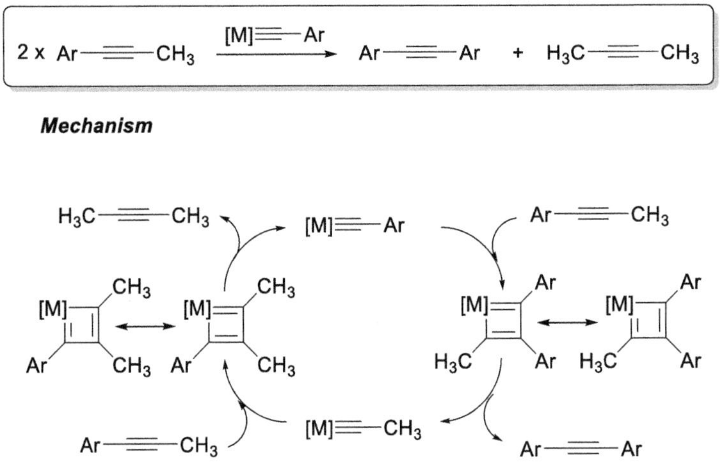

Figure 2.26 A general mechanism for alkyne metathesis reactions.

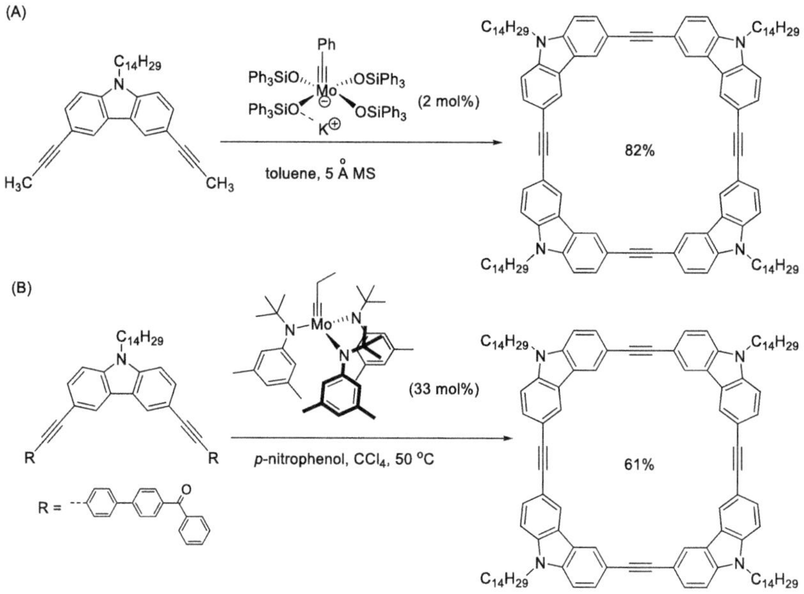

Figure 2.27 Two alkyne metathesis approaches to generate a carbazole-ethynylene shape-persistent macrocycle.

different alkyne metathesis approaches [25]. The first approach (Figure 2.27A) uses a methyl-capped alkyne precursor as well as molecular sieves to trap the butyne by-product. In the other approach (Figure 2.27B), the precursor dialkyne is designed with certain phenyl ketone end groups. After metathesis one of the alkyne products precipitates out the solution due to its low solubility. The precipitation in turn pushes the reaction equilibrium in favor of the macrocyclization direction.

The remarkable efficiency of alkyne metathesis in macrocyclization was show-cased by a high-yielding synthesis of a Möbius tris((ethynyl)[5]helicene) macro-cycle reported by Moore and co-workers in 2020 (Figure 2.28) [26]. This one-pot macrocyclization has achieved a stunningly high yield of 86% starting from a dialkynyl-substituted [5]helicene precursor.

The preparation of π-conjugated polymers can be conducted through various alkyne metathesis polymerization reactions; for example, ring-opening alkyne polymerization and acylic diyne metathesis polymerization (ADMET). It is worth noting that the ADMET strategy (see Figure 2.29) has attracted growing interest in recent years [27, 28]. Compared with traditional cross-coupling methodologies such as Sonogashira coupling, this approach can effectively avoid undesired homocoupling side reactions, hence resulting in polymer products with fewer structural defects. Moreover, ADMET reaction conditions can be tuned to give high molecular weight polymers with controllable polydispersity, boding well for the application of the ADMET strategy to alkyne-based π-conjugated organic polymers and nanomaterials.

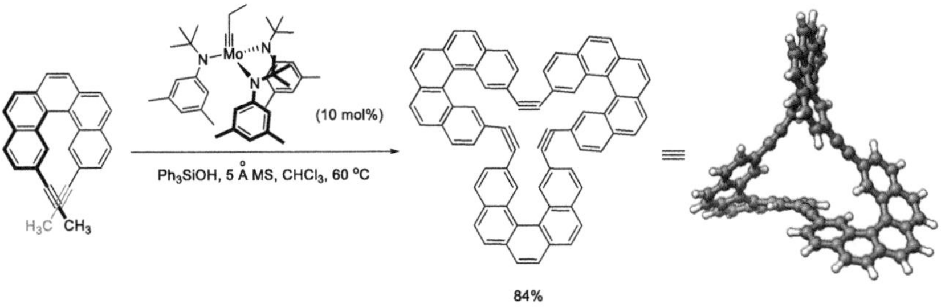

84%

Figure 2.28 Synthesis of a tris((ethynyl)[5]helicene) macrocycle with a Möbius topology through alkyne metathesis.

Figure 2.29 General synthetic strategy for making π-conjugated poly(arylene ethynylene)s through acyclic diyne metathesis polymerization (ADMET).

2.3.5 Click Reactions

Assembly of molecular building blocks together through covalent and non-covalent linkages constitutes the main task in the design and preparation of organic nanomaterials. The construction of covalent linkages can be done through specific organic reactions such as the C–C bond forming reactions discussed above. However, the application of C–C bond forming reactions to the preparation of nanomaterial is strongly dependent on the scope of the reactions involved as well as their tolerance to other materials and functional groups present during the synthesis. When dealing with a large number of synthetic targets, for example making a library of analogues for later studies of structure–property relationships, traditional synthetic approaches are insufficient in dealing with the challenges involved. High yielding and broad scope synthetic methodologies that can efficiently join (or "click") two or more molecular units together are hence highly desirable. Nature produces complicated biological molecules (e.g., proteins, polysaccharides) through a "modular" approach, where simple units are covalently linked together by certain carbon–heteroatom bonds. Motivated by this approach, an innovative synthetic concept, called **click chemistry**, was conceived in the late 1990s and then rapidly developed. The term "click chemistry" was first coined in 1999 by K. B. Sharpless, a renowned synthetic organic chemist who won the 2001 Nobel Prize in Chemistry for his pioneering work on chirally catalyzed oxidation reactions. Sharpless introduced the concept of click chemistry at the 217th American Chemical Society annual meeting, and it immediately became a highly popular topic. Later, in his landmark review in 2001, Dr. Sharpless clearly defined click chemistry as a group of reactions that must be modular, wide in scope, give very high yields, generate only inoffensive by-products that can be removed by nonchromatographic methods, and be stereospecific (but not necessarily enantioselective) [29]. In 2001, Meldal and Sharpless independently reported a Cu-catalyzed azide alkyne coupling reaction, which is nowadays well known as the CuAAC reaction [30]. The discovery of this reaction provided a huge incentive for the development and application of click chemistry.

Click chemistry is also known as linkage chemistry, dynamic chemistry, combinatorial chemistry, or quick linking combinatorial chemistry. Ever since its inception, click chemistry has undergone enormous growth and development. As a result, it has become one of the most useful synthetic strategies in many fields [31, 32]. It is worth noting that unlike the abovementioned transition metal-mediated reactions, click chemistry is a very successful concept applied to organic synthesis rather than a special class of organic reactions. In addition to the CuAAC reaction, which has been widely regarded as the flagship click reaction, a wide range of organic reactions can also be considered as choices for implementing click chemistry, so long as they possess the following characteristics.

1) Click reactions use raw materials and reagents that are abundantly available and inexpensive.
2) Click reactions mainly lead to the formation of carbon–heteroatom bonds (e.g., C–O, C–N, and C–S).
3) Click reactions are simply performed under mild conditions without significant perturbation by air and water.
4) Click reactions are high yielding and highly stereoselective.
5) Click reactions are exothermic transformations from high-energy reactants to stable products. According to the Sharpless' definition, a click reaction is defined by a gain of thermodynamic enthalpy (ΔH_r) of at least 20 kcal mol^{-1}. This feature ensures that a click reaction will be high yielding and nearly substrate-insensitive.
6) Click reactions usually produce no by-products or only water as a nontoxic by-product.
7) The products of click reactions are easy to purify, for example, through crystallization and distillation. Complex chromatographic separations should be avoided.

According to the criteria discussed above, a wide range of organic reactions have been recognized as useful "click" ligation methods for linking different molecular fragments together or modularly functionalizing the backbones of polymers, biomolecules, and the surfaces of nanoparticles and self-assembled monolayers (SAMs). Figure 2.30 illustrates a collection of popularly used click reactions, in which the CuAAC reaction has been widely deemed as the flagship reaction of the click chemistry arsenal, owing to its remarkable advantages of

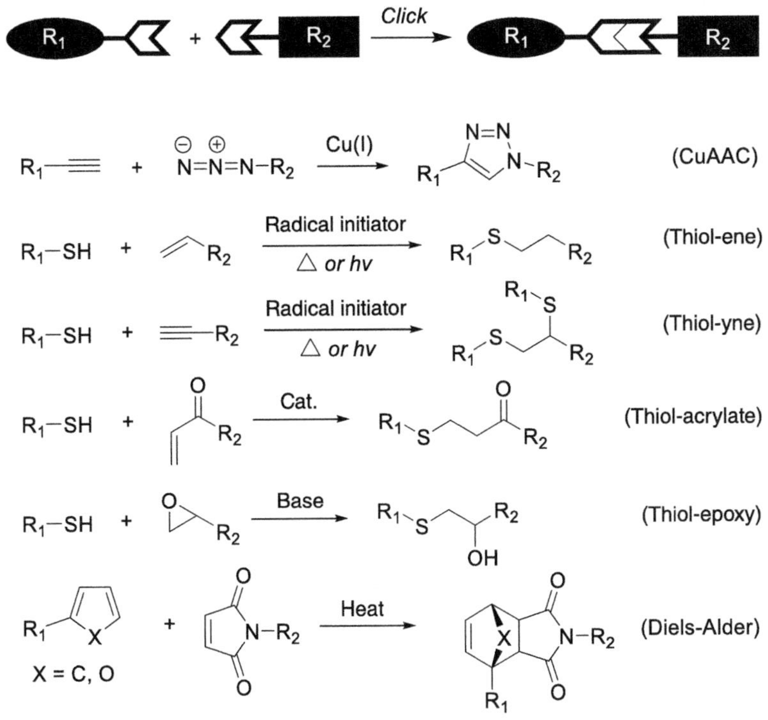

Figure 2.30 A general scheme of click chemistry and commonly used click reactions.

straightforward procedures, high functional group tolerance, and usually high yields. The CuAAC reaction is a modified Hüisgen [3+2] cycloaddition that is catalyzed by Cu(I). The uncatalyzed Hüisgen reaction follows a concerted 1,3-dipolar cycloaddition mechanism, resulting in a mixture of 1,4- and 1,5-disubstituted regioisomers as the products. The CuAAC reactions, however, follow a stepwise mechanism and exclusively produce 1,4-disubstituted triazoles as the products. The reactions usually afford very high yields and show excellent tolerance to water. The catalytic Cu(I) species can be generated by various methods, including (i) direct utilization of a Cu(I) source, (ii) reduction of a Cu(II) source, and (iii) the oxidation of elemental Cu(0) to form Cu(I). For example, the early version of the catalytic system devised by Sharpless and Folkin used a mixture of $CuSO_4$ and sodium ascorbate to generate Cu(I) *in situ* in the presence of water [33]. Later on, a wide range of Cu catalysts, such as phosphine-based Cu(I) complexes, N-heterocyclic carbene (NHC)-based Cu(I) complexes, and Cu nanoparticles with excellent catalytic activities, offered a wider scope of reaction conditions to choose from in order to carry out CuAAC reactions in an efficient, selective, and cost-effective manner.

Numerous mechanisms for the CuAAC reactions have been proposed by various research groups including Sharpless himself. All the mechanisms were based on experimental results and/or computational studies (e.g., DFT calculations) and generally agreed with the experimental observations made in the CuAAC reactions. The most conclusive mechanism so far has been the one reported by Bertrand and co-workers [34], who successfully isolated and characterized a biscopper acetylide complex, **I(Cu₂)**, and a 3,5-bismetallated triazole complex, **II(Cu₂)**, as the active intermediates in a kinetically favored catalytic cycle (see Figure 2.31).

Azide-alkyne cycloaddition reactions can also be catalyzed by other transition metals to achieve "click-type" functionalization. For example, Ru-catalyzed azide-alkyne cycloaddition (RuAAC) was first reported in 2005, showing a regioselectivity for the 1,5-disubstituted 1,2,3-triazole product [35]. RuAAC reactions show tolerance to a range of functional groups, but they suffer from high sensitivity toward solvents and the steric demand of azide substituents [36]. Therefore, the RuACC reactions have been explored to a lesser degree than the CuAAC reactions in click chemistry. Ag-catalyzed azide-alkyne coupling (AgAAC) has also been reported, showing selectivity for 1,4-disubstituted triazoles through cycloaddition. However, this method is also not as popular as the CuAAC reactions due to the high cost of silver catalysts. Other transition metals, such as Zn and Pd, have been investigated as well, but their application to azide–alkyne click reactions has been limited.

Without any doubt, the CuAAC reaction is one of the most powerful and therefore most popular method for modular functionalization of various organic and biomaterials. Many CuAAC reactions can be readily performed in homogeneous phases or at various interfaces, owing to their high yields, excellent selectivity, and tolerance to reactions conditions. Figure 2.32 shows an example of modular functionalization of a [60]fullerene hexaadduct using the CuSO₄/sodium ascorbate catalyzed azide–alkyne coupling reaction, reported by Nierengarten and co-workers [37]. Herein, the click functionalization approach successfully attached six arene end groups, ranging from simple benzenes to a relatively large porphyrin derivative, to the fullerene core with good to excellent yields. CuAAC reactions can also be performed on the surface of carbon nanomaterials, for example, single-walled carbon nanotubes, to have a variety of functional molecular or macromolecular moieties installed on it. Adronov and co-workers demonstrated the efficiency of this approach by "clicking" azido-capped poly(styrene)s onto the sidewall of single-walled carbon nanotubes (Figure 2.33) [38].

Click reactions can be applied to self-assembled monolayers (SAMs) to achieve controlled patterning and functionalization. Figure 2.34 illustrates the work by Reinhoudt and co-workers, who employed click chemistry to modify an alkyl azide SAM on a silicon oxide substrate [39]. A special technique called microcontact print was used to transfer patterns onto the SAM surface. As illustrated in Figure 2.34A, an azido-terminated alkyl SAM was first prepared on the surface of

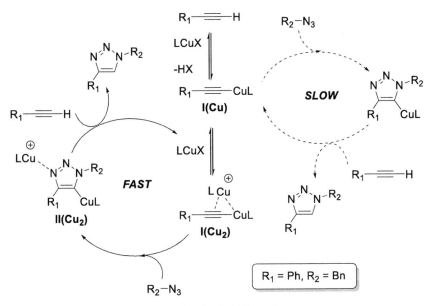

Figure 2.31 Bertrand's mechanism for the CuAAC reactions.

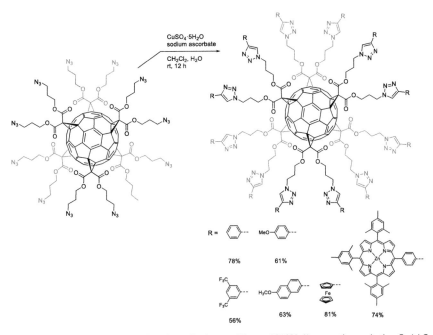

Figure 2.32 Click functionalization of a hexaadduct of [60]fullerene through the CuAAC reaction.

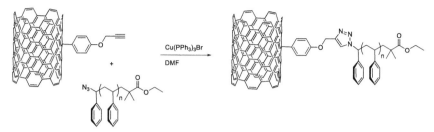

Figure 2.33 Click functionalization of single-walled carbon nanotubes with poly(stryrene)s through the CuAAC reaction.

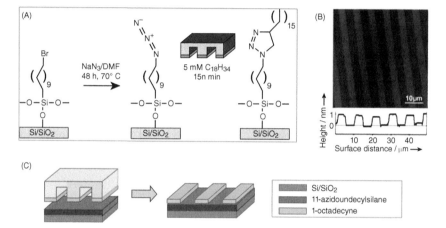

Figure 2.34 (A) Click functionalization of an alkyl azide SAM on a silicon oxide substrate through microcontact printing. (B) Atomic force microscopic (AFM) image of the line pattern obtained by printing 1-octadecyne onto the azido-terminated SAM. (C) Schematic illustration of the microcontact printing method. (Reproduced from Angew. Chem. Int. Ed. 2006 / Rozkiewicz et al.,2006 / JOHN WILEY & SONS, INC.)

a silicon oxide substrate. Then a patterned stamp loaded with 1-octadecyne was applied to the SAM surface. Within a contact time of 15 min, azide–alkyne cyclo-addition reactions occurred in the contact regions. This method was reported to be highly efficient and did not require the presence of any Cu catalysts.

In 2004, an elegantly designed catalyst-free azide-alkyne cycloaddition was reported by Bertozzi and co-workers, taking advantage of the strain energy stored in cycloalkynes as the driving force for "click functionalization" [40]. For example, the alkyne bond angle in cyclooctyne is deformed to 163°, offering a strain energy of 18 kcal mol^{-1} toward the induction of a spontaneous cycloaddition with an azide partner (Figure 2.35). This type of reaction has been well known as the **strain-promoted azide–alkyne cycloaddition** (SPAAC) [41]. Compared with the CuAAC reaction, the SPAAC process does not require the participation of toxic Cu catalysts; therefore, it is particularly useful for connecting tags or labels to biological molecules such as proteins and DNA.

The numerous types of click reactions developed over the past two decades have greatly expanded the applicability of click chemistry to not only organic and biological nanomaterials but also living organisms. Indeed, click chemistry has been demonstrated to proceed fast and reliably in complex living organisms that contain a range of biofunctionalities. Furthermore, click functionalization neither perturbs the native activities of the living organisms nor generates cytotoxic by-products. This type of functionalization, known as the bio-orthogonal click chemistry, has been applied with great success in bioengineering, therapeutic and diagnostic technologies. Along this direction, continuing efforts are being made toward the development of new click reactions. In 2022 the Nobel Prize in Chemistry was awarded to Carolyn R. Bertozzi, Morten Meldal, and K. Barry Sharpless for their pioneering contributions to the development of click chemistry and bioorthogonal chemistry. For Sharpless, this is his second reception of the Nobel Prize.

Figure 2.36 shows an example of functionalization of mammalian cells using a bio-orthogonal click strategy [42]. At first a simple alkene tag, homoallylglycine, was co-translationally incorporated onto the proteins in cells as a clickable synthetic handle and then subjected to a photoinduced cycloaddition (click) reaction with a diaryltetrazole for 5 seconds using a femtosecond 700 nm laser. This photoclick method can be applied to achieve *in vitro* functionalization and spatio-temporally controlled imaging.

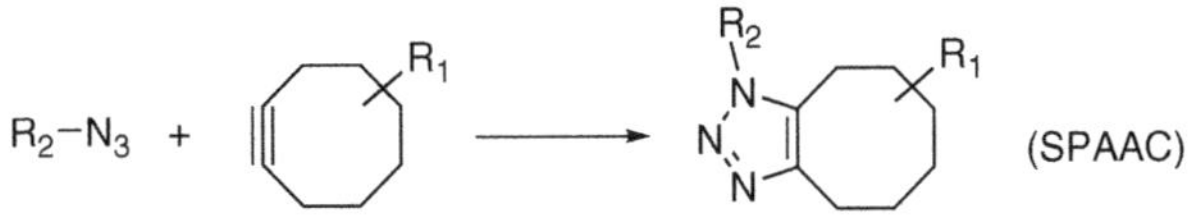

Figure 2.35 General reaction scheme for strain-promoted azide–alkyne coupling (SPAAC) reactions.

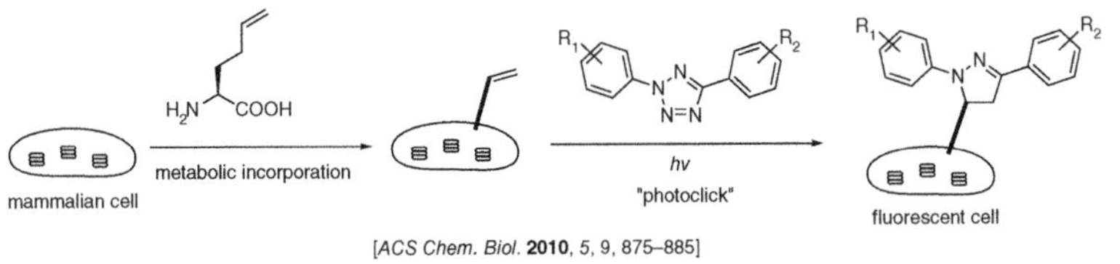

[*ACS Chem. Biol.* **2010**, 5, 9, 875–885]

Figure 2.36 Bio-orthogonal click functionalization of mammalian cells through a photoinduced tetrazole–alkene cycloaddition reaction.

2.3.6 Dynamic Covalent Chemistry

Dynamic covalent chemistry (DCC) is a synthetic concept particularly useful for constructing highly organized, discrete nanoarchitectures and supramolecular systems [43–45]. DCC utilizes a variety of reversible reactions performed under thermodynamic control. The reversible nature of the reactions allows "error checking" and "proof-reading" to drive the synthesis toward the most stable product under conditions being used. Numerous thermodynamic, kinetic, and experimental parameters come into play during the dynamic covalent synthesis. These have offered abundant opportunities for chemists to design and prepare complex structures, ranging from macrocycles, molecular cages, to 2D and 3D molecular frameworks.

Actually, some TM-catalyzed macrocyclization reactions illustrated in the previous sections are based on dynamic covalent reactions; for example, the olefin metathesis reaction shown in Figure 2.25A and alkyne metathesis reactions shown in Figure 2.27. Other types of dynamic covalent reactions have also been well established, which mainly fall into the following classes: (i) exchange reactions, (2) condensation reactions (e.g., imine condensation), and (3) addition reactions (e.g., aldol reactions, cycloadditions). The dynamic bonds in DCC encompass a wide range, including C–C, C–O, C–N C–S, B–O, and S–S bonds.

Figure 2.37 shows the classical synthesis of calix[4]arenes through the reaction of 4-*tert*-butylphenol with formaldehyde reported by Gutsche in 1990 [46]. A **calix[*n*]arene** is a macrocycle based on methylene (–CH_2–) linked phenols. The nomenclature consists of two parts. "Calix" is the Greek name for chalice, in view of the resemblance of this type of molecule to a vase (or cup) in shape, while "arene" refers to the aromatic building blocks and *n* is the number of repeat units in the cyclic array. The formation of calix[4]arene takes place through a reversible process that is under thermodynamic control. Owing to their unique molecular shapes, calixarenes serve as an important class of supramolecular hosts to trap various molecular and ionic guests into their hydrophobic π-cavities.

Analogous to the calix[*n*]arenes is a class of cyclic compounds named **pillar[*n*] arenes**, which can be prepared through a reversible Friedel-Crafts alkylation reaction. Figure 2.38 illustrates the synthesis of pillar[5]arene from the Fridel-Crafts

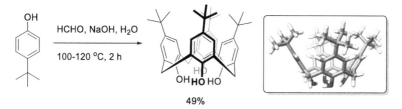

Figure 2.37 Gutsche's synthesis of calix[4]arene. Inset: molecular model of calix[4]arene in its stable conformation.

reaction between 2,5-bis(bromomethyl)-1,4-diethoxybenzene and 1,4-diethoxy-benzene reported by Nierengarten and co-workers [47]. Ideally, the reaction should give a macrocycle with an even number of arene repeat units, especially a hexameric cyclic oligomer (i.e., pillar[6]arene) which was conceived as a possible product. Experimentally, only the thermodynamically stable pillar[5]arene was obtained in a good yield from this reaction, attesting to the thermodynamically controlled nature of this reaction.

In recent years, dynamic covalent reactions have found extensive application in the generation of organized 2D and 3D molecular frameworks, such as **covalent organic frameworks** (COFs) [48–50]. COFs are a class of pre-designable polymers that are assembled through topology-diagram-directed polymer growth along with geometry matching of monomeric units. The conformations and morphologies of COFs create nanoscale molecular spaces and interfaces that provide special functions for various applications. Figure 2.39 shows the synthesis of a 2D-COF through the imine condensation between an amino and an aldehyde-substituted tetraphenylethene (TPE) building block [51].

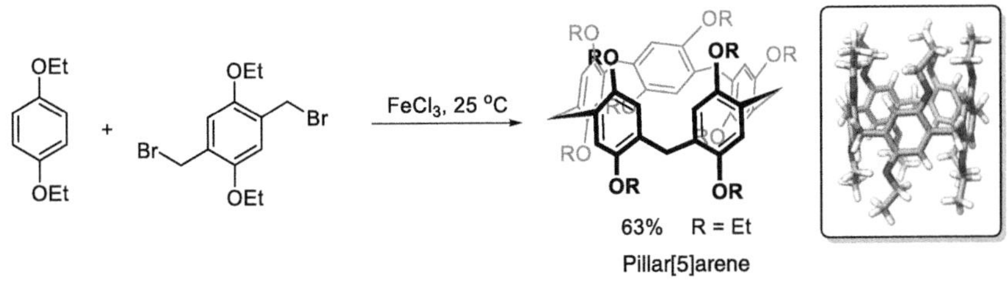

Figure 2.38 Thermodynamically controlled synthesis of pillar[5]arene from a reversible Friedel-Crafts alkylation approach. Inset: molecular model of pillar[5]arene.

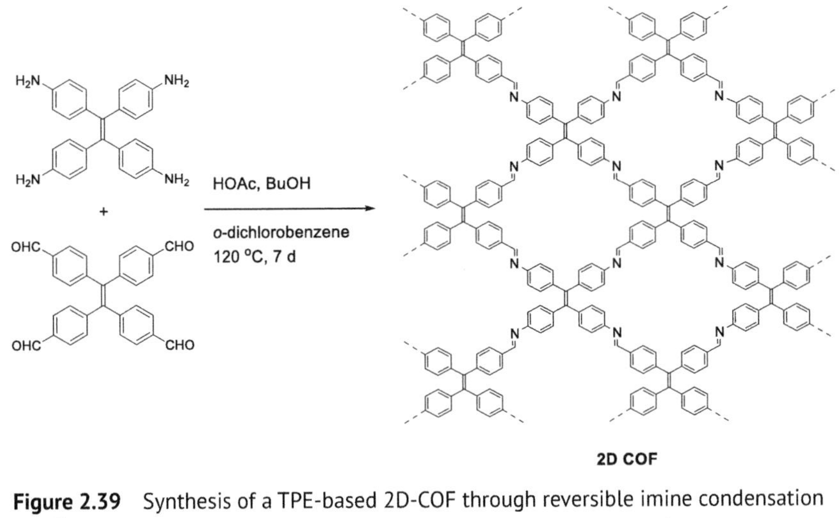

Figure 2.39 Synthesis of a TPE-based 2D-COF through reversible imine condensation reactions.

2.4 Methods for Surface Functionalization

Surface chemistry has a long well-documented history. However, its development underwent rapid expansion only after the emergence of nanotechnology in the late 1970s, when active research activity became focused on the derivatization of surfaces with functional molecules and polymers. In the 1980s, alkanethiols were found to spontaneously assemble on noble metals such as gold. This exciting discovery opened the door to creating surfaces for virtually any desired chemistry by placing a gold substrate into a millimolar solution of an alkanethiol in ethanol. The crystalline-like monolayers formed on the metal surface by this approach are called self-assembled monolayers (SAMs) [52].

Gold is a versatile metal allowing SAMs to be readily attached onto its surface. SAMs are an important element of modern nanotechnology, which are spontaneously formed on solid substrates through covalent and non-covalent attachments. SAM-coated gold surfaces can be synthesized to possess special electronic and optoelectronic properties, leading to a wide range of applications in nanotechnology. Moreover, the formation of SAMs can stabilize gold nanoparticles, which in turn serve as versatile nanoscale platforms for applications in chemical sensing, bioimaging, drug delivery, therapeutics and diagnostics, energy conversion, catalysis, just to name a few.

Typical alkanethiol molecules interact with a gold surface through chemisorption (see Figure 2.40). On a flat gold surface, alkanethiol chains are arranged with a tilt angle of about 30° in the monolayer. Each alkanethiol molecule is bound to a gold atom, forming a semi-covalent bond with a bond strength around 45 kcal mol^{-1}. The van der Waals interactions and hydrophobicity of the alkyl chains provide another driving force for the formation of alkanethiol SAMs. Moreover, the end of the alkyl chain can be flexibly modified with various types of functional

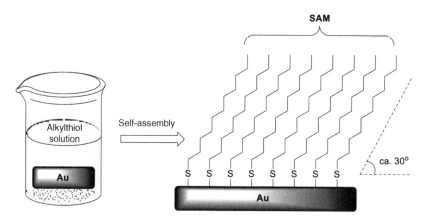

Figure 2.40 Generation of an alkanethiol SAM on gold surface.

groups (e.g., hydrophilic groups, synthetic receptors, biomolecules) to incorporate new functionalities onto alkanethiol SAMs.

A free alkylthiol group may incur some challenges in multiple-step organic synthesis due to its acidity and reactivity. To avoid such problems, a thiol group can be first protected as an acetylthiol. At the stage of SAM formation, hydrolysis of the acetylthiol can be induced under mild conditions (e.g., treatment with cyanide anion) to give the free thiol group (Figure 2.41A). Similarly, an alkanethiol SAM can be formed through the cleavage of a disulfide bond in a cyclic disulfide (Figure 2.41B). This type of anchor group leads to enhanced surface attachment due to the formation of two Au–S bonds.

Tripodal ligands are advantageous anchor groups for assembling functional molecules on surfaces (e.g., a gold substrate or a nanoparticle), since they not only form strong linkages (three Au–S bond per molecule), but hold the functionalities in an upright orientation for better surface performance. An elegant example of these is illustrated in Figure 2.42. In this work, Feringa and co-workers used a tripodal ligand to assemble an array of photo-driven molecular motors on the gold surface [53]. The detailed mechanism for the rotatory performance of such photo-driven molecular motors will be discussed in Chapter 5. The photo-driven molecular motor is functionalized with a hydrophobic perfluoroalkyl unit. As such, the rotation of the motor modifies the wettability of the SAM-coated surface. It is worth remarking that Prof. Ben L. Feringa is well-known for his ingenious work on molecular motors and molecular machines. His pioneering work in this field led him to winning of the 2016 Nobel Prize in Chemistry, which he shared with two other renowned chemists, Fraser Stoddart and Jean-Pierre Sauvage.

Besides sulfur-based linkages, SAMs on gold and other metal surfaces can be constructed using other types of ligands such as carboxylic and phosphine groups. In 1991, Arduengo and co-workers isolated and characterized the first stable *N*-heterocyclic carbene (NHC), which immediately attracted enormous interest in the development of NHC derivatives within the synthetic community [54]. Many applications of NHCs are focused on their use as active ligands for organic catalysis [55, 56]. As shown in Figure 2.43, a stable NHC has a carbene center with a filled orbital and an empty orbital. These orbitals allow it to interact with the *d* orbitals of a transition metal in different modes, among which the σ-donation,

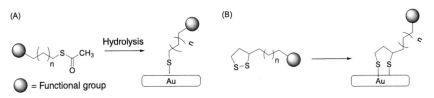

Figure 2.41 Formation of SAMs on gold surface through (A) acetylthiol and (B) disulfide anchor groups.

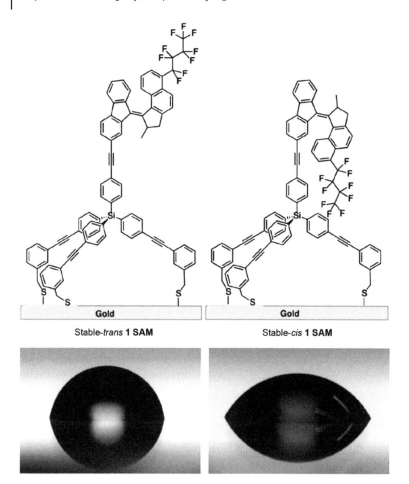

Stable-*trans* **1 SAM**   Stable-*cis* **1 SAM**

Figure 2.42 Top: immobilization of a photo-driven molecular motor on a gold surface through a tripodal ligand. Bottom: photographed images of a water droplet on the surfaces of a molecular motor SAM in different orientations. (Reproduced from J. Am. Chem. Soc. 2014 / Ivashenko et al.,2014 / American Chemical Society.)

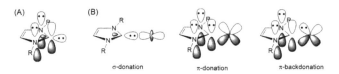

Figure 2.43 (A) Structure and orbital properties of a stable NHC. (B) The most important interaction modes between an NHC and a transition metal.

(A)

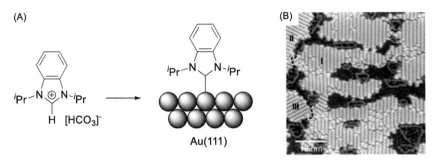

(B)

Figure 2.44 (A) Formation of an NHC SAM on an Au(111) surface. (B) STM images of the NHC SAM. (Reproduced from *Nat. Commun.* 2021, *12*, 4034.)

π-donation, and π-backdonation are the most important ones. Crudden and Horton in 2014 demonstrated that NHCs can form highly stable SAMs on a gold surface [57]. Since then, NHCs have been extensively investigated as appealing alternatives to thiols for the formation of robust and ordered SAMs on transition metal surfaces. NHCs have been found to form SAMs with exceptionally high stability. The resulting NHC-based SAMs are clean and organized, tolerating a wide range of conditions, such as boiling organic solvents, acids, bases, and oxidants, which would otherwise destroy typical thiol SAMs on gold. Figure 2.44 illustrates an example of NHC-based SAMs on an Au(111) surface recently reported by Crudden and co-workers [58].

In addition to the metal ligands mentioned above, the functionalization of a surface can be achieved by the treatment of an arene diazonium salt. Arene diazonium ions are useful reactive intermediates in classical organic synthesis, for example, the Sandmeyer reaction, the diazo coupling, and the Schiemann-Baltz reaction. Electrochemical reduction of an arene diazonium salt can lead to surface functionalization (electrografting) of a wide range of substrates, including carbon, metal, and semiconductor materials. As shown in Figure 2.45, the electrografting of a diazonium salt involves the formation of an aryl radical, which is subsequently bonded to the surface through electron transfer. Diazonium salts can be readily prepared, but most of them are explosive when dry or at room temperature or above. For safety reasons, diazonium tetrafluoroborates ($ArN_2^+ \cdot BF_4^-$) are preferred as they are stable and do not cause an explosion upon heating. Owing to the high reactivity of aryl radicals, it is possible for them to react with the arenes already grafted to the surface, resulting in complex polyarene structures. For certain types of surface functionalization tasks, this may not be a desired scenario.

Arene diazonium salts are extremely usefully anchoring groups to link various organic functional molecules/macromolecules to carbon surfaces, for example, highly ordered pyrolytic graphite (HOPG), single-walled carbon nanotubes (SWCNTs), graphenes, and carbon nanodots. Besides electrochemical means, the surface

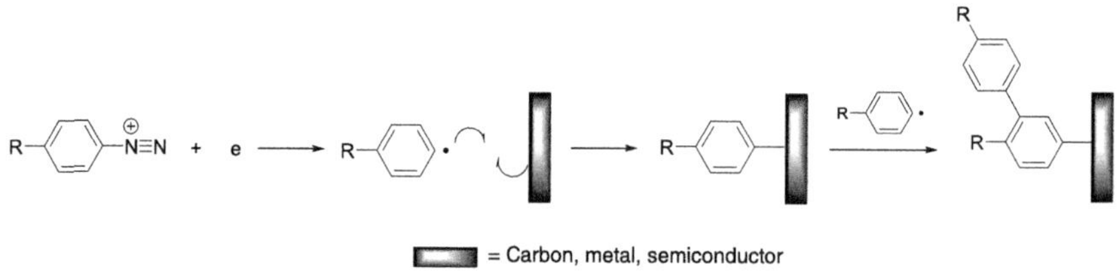

Figure 2.45 Mechanism for electrografting of an arene diazonium ion on a surface.

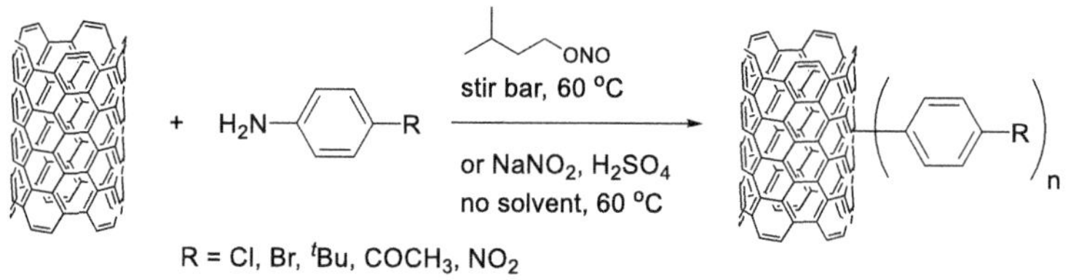

R = Cl, Br, ^{t}Bu, COCH$_3$, NO$_2$

Figure 2.46 Covalent functionalization of single-walled carbon nanotubes with diazonium salts under solvent-free conditions.

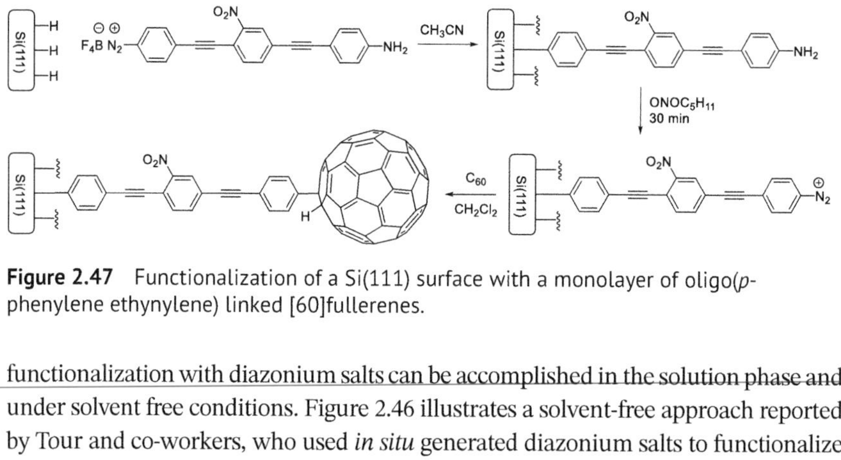

Figure 2.47 Functionalization of a Si(111) surface with a monolayer of oligo(*p*-phenylene ethynylene) linked [60]fullerenes.

functionalization with diazonium salts can be accomplished in the solution phase and under solvent free conditions. Figure 2.46 illustrates a solvent-free approach reported by Tour and co-workers, who used *in situ* generated diazonium salts to functionalize the sidewall of single-walled carbon nanotubes with a range of aryl substituents [59].

Finally, it is worth mentioning that crystalline silicon (e.g., Si(111)), which under ambient conditions is covered by a layer of SiO$_2$, can be converted into a monocrystalline surface of hydrogenated silicon. This hydrogenated silicon surface can be electrografted with functionalization by diazonium salts, which offers an efficient approach to modify semiconducting silicon surface for various electronic applications. Figure 2.47 shows a stepwise surface functionalization of hydrogenated Si(111) wafer with a monolayer of [60]fullerene molecules covalently linked to the surface through an oligo(*p*-phenylene ethynylene) bridge. In this synthesis, diazonium salts are used to effectively generate covalent linkage on the Si(111) surface as well as to the cage of fullerene [60].

Further Reading

- Carey, F. A.; Sundberg, R. J., Advanced Organic Chemistry: Part B: Reactions and Synthesis. Springer: 2007.
- Warren, S.; Wyatt, P., Organic Synthesis: the Disconnection Approach. 2nd ed.; John Wiley & Sons: 2008.
- De Meijere, A.; Bräse, S.; Oestreich, M., Metal-Catalyzed Cross-Coupling Reactions and More. Wiley-VCH: Weinheim, 2014.

References

1 Colacot, T. J., New Trends in Cross-Coupling: Theory and Applications. Royal Society of Chemistry: Cambridge, UK, **2014**.
2 Tour, J. M., Conjugated Macromolecules of Precise Length and Constitution. Organic Synthesis for the Construction of Nanoarchitectures. *Chem. Rev.* **1996**, *96*, 537–554.
3 Schumm, J. S.; Pearson, D. L.; Tour, J. M., Iterative Divergent/Convergent Approach to Linear Conjugated Oligomers by Successive Doubling of the Molecular Length: A Rapid Route to A 128Å-Long Potential Molecular Wire. *Angew. Chem. Int. Ed. Engl.* **1994**, *33*, 1360–1363.
4 Davis, J. J.; Coleman, K. S.; Busuttil, K. L.; Bagshaw, C. B., Spatially Resolved Suzuki Coupling Reaction Initiated and Controlled Using a Catalytic AFM Probe. *J. Am. Chem. Soc.* **2005**, *127*, 13082–13083.
5 Alberico, D.; Scott, M. E.; Lautens, M., Aryl–Aryl Bond Formation by Transition-Metal-Catalyzed Direct Arylation. *Chem. Rev.* **2007**, *107*, 174–238.
6 McGlacken, G. P.; Bateman, L. M., Recent Advances in Aryl–Aryl Bond Formation by Direct Arylation. *Chem. Soc. Rev.* **2009**, *38*, 2447–2464.
7 Liu, S.-Y.; Cheng, J.-Z.; Zhang, X.-F.; Liu, H.; Shen, Z.-Q.; Wen, H.-R., Single-Step Access to a Series of D–A π-Conjugated Oligomers with 3–10 nm Chain Lengths. *Polym. Chem.* **2019**, *10*, 325–330.
8 Hay, A. S., Oxidative Coupling of Acetylenes. II1. *J. Org. Chem.* **1962**, *27*, 3320–3321.
9 Bohlmann, F.; Schönowsky, H.; Inhoffen, E.; Grau, G., Polyacetylenverbindungen, LII. Über den Mechanismus der Oxydativen Dimerisierung von Acetylenverbindungen. *Chem. Ber.* **1964**, *97*, 794–800.
10 Fischer, M.; Höger, S., Synthesis of a Shape-Persistent Macrocycle with Intraannular Carboxylic Acid Groups. *Tetrahedron* **2003**, *59*, 9441–9446.
11 Movsisyan, L. D.; Franz, M.; Hampel, F.; Thompson, A. L.; Tykwinski, R. R.; Anderson, H. L., Polyyne Rotaxanes: Stabilization by Encapsulation. *J. Am. Chem. Soc.* **2016**, *138*, 1366–1376.

12 Venkataramana, G.; Dongare, P.; Dawe, L. N.; Thompson, D. W.; Zhao, Y.; Bodwell, G. J., 1, 8-Pyrenylene–Ethynylene Macrocycles. *Org. Lett.* **2011**, *13*, 2240–2243.

13 Liang, S.; Chen, G.; Zhao, Y., Conformationally Switchable TTFV–Phenylacetylene Polymers: Synthesis, Properties, and Supramolecular Interactions with Single-Walled Carbon Nanotubes. *J. Mater. Chem. C* **2013**, *1*, 5477–5490.

14 Sato, H.; Bender, J. A.; Roberts, S. T.; Krische, M. J., Helical Rod-Like Phenylene Cages via Ruthenium Catalyzed Diol-Diene Benzannulation: A Cord of Three Strands. *J. Am. Chem. Soc.* **2018**, *140*, 2455–2459.

15 Fürstner, A., *Alkene Metathesis in Organic Synthesis.* Springer-Verlag: Berlin, **2003**.

16 Karabulut, S., *Metathesis Chemistry: From Nanostructure Design to Synthesis of Advanced Materials.* Springer: Dordrecht, **2007**.

17 Grela, K., *Olefin Metathesis: Theory and Practice.* John Wiley & Sons: Hoboken, **2014**.

18 Jin, Y.; Zhang, A.; Huang, Y.; Zhang, W., Shape-Persistent Arylenevinylene Macrocycles (AVM) Prepared via Acyclic Diene Metathesis Macrocyclization (ADMAC). *Chem. Commun.* **2010**, *46*, 8258–8260.

19 Zhang, Q.; Zhang, Y.-E.; Tong, S.; Wang, M.-X., Hydrocarbon Belts with Truncated Cone Structures. *J. Am. Chem. Soc.* **2020**, *142*, 1196–1199.

20 Chang, S.-W.; Horie, M., A donor–Acceptor Conjugated Block Copolymer of Poly(arylenevinylene)s by Ring-Opening Metathesis Polymerization. *Chem. Commun.* **2015**, *51*, 9113–9116.

21 Fürstner, A.; Davies, P. W., Alkyne Metathesis. *Chem. Commun.* **2005**, 2307–2320.

22 Fürstner, A., Alkyne Metathesis on the Rise. *Angew. Chem. Int. Ed.* **2013**, *52*, 2794–2819.

23 Zhang, W.; Moore, J. S., Alkyne Metathesis: Catalysts and Synthetic Applications. *Adv. Synth. Catal.* **2007**, *349*, 93–120.

24 Mortreux, A.; Blanchard, M., Metathesis of Alkynes by a Molybdenum Hexacarbonyl–Resorcinol Catalyst. *J. Chem. Soc., Chem. Commun.* *1974*, 786–787.

25 Zhang, W.; Moore, J. S., Arylene Ethynylene Macrocycles Prepared by Precipitation-Driven Alkyne Metathesis. *J. Am. Chem. Soc.* **2004**, *126*, 12796-12796.

26 Jiang, X.; Laffoon, J. D.; Chen, D.; Pérez-Estrada, S.; Danis, A. S.; Rodríguez-López, J.; Garcia-Garibay, M. A.; Zhu, J.; Moore, J. S., Kinetic Control in the Synthesis of a Möbius Tris((ethynyl)[5] helicene) Macrocycle Using Alkyne Metathesis. *J. Am. Chem. Soc.* **2020**, *142*, 6493–6498.

27 Mutlu, H.; de Espinosa, L. M.; Meier, M. A. R., Acyclic Diene Metathesis: A Versatile Tool for the Construction of Defined Polymer Architectures. *Chem. Soc. Rev.* **2011**, *40*, 1404–1445.

28 Caire da Silva, L.; Rojas, G.; Schulz, M. D.; Wagener, K. B., Acyclic Diene Metathesis Polymerization: History, Methods and Applications. *Prog. Polym. Sci.* **2017**, *69*, 79–107.

29 Kolb, H. C.; Finn, M. G.; Sharpless, K. B., Click Chemistry: Diverse Chemical Function from a Few Good Reactions. *Angew. Chem. Int. Ed.* **2001**, *40*, 2004–2021.

30 Hein, J. E.; Fokin, V. V., Copper-Catalyzed Azide–Alkyne Cycloaddition (CuAAC) and Beyond: New Reactivity of Copper(i) Acetylides. *Chem. Soc. Rev.* **2010**, *39*, 1302–1315.

31 Lahann, J., Click Chemistry: A Universal Ligation Strategy for Biotechnology and Materials Science. In *Click Chemistry for Biotechnology and Materials Science*, Lahann, J., Ed. John Wiley & Sons: **2009**; pp 1–7.

32 Moses, J. E.; Moorhouse, A. D., The Growing Applications of Click Chemistry. *Chem. Soc. Rev.* **2007**, *36*, 1249–1262.

33 Rostovtsev, V. V.; Green, L. G.; Fokin, V. V.; Sharpless, K. B., A Stepwise Huisgen Cycloaddition Process: Copper(I)-Catalyzed Regioselective "Ligation" of Azides and Terminal Alkynes. *Angew. Chem. Int. Ed.* **2002**, *41*, 2596–2599.

34 Jin, L.; Tolentino, D. R.; Melaimi, M.; Bertrand, G., Isolation of Bis(copper) Key Intermediates in Cu-Catalyzed Azide-Alkyne "Click Reaction". *Sci. Adv.* **2015**, *1*, e1500304.

35 Zhang, L.; Chen, X.; Xue, P.; Sun, H. H. Y.; Williams, I. D.; Sharpless, K. B.; Fokin, V. V.; Jia, G., Ruthenium-Catalyzed Cycloaddition of Alkynes and Organic Azides. *J. Am. Chem. Soc.* **2005**, *127*, 15998–15999.

36 Johansson, J. R.; Beke-Somfai, T.; Said Stålsmeden, A.; Kann, N., Ruthenium-Catalyzed Azide Alkyne Cycloaddition Reaction: Scope, Mechanism, and Applications. *Chem. Rev.* **2016**, *116*, 14726–14768.

37 Iehl, J.; Pereira de Freitas, R.; Delavaux-Nicot, B.; Nierengarten, J.-F., Click Chemistry for the Efficient Preparation of Functionalized [60]fullerene Hexakis-Adducts. *Chem. Commun.* **2008**, 2450–2452.

38 Li, H.; Cheng, F.; Duft, A. M.; Adronov, A., Functionalization of Single-Walled Carbon Nanotubes with Well-Defined Polystyrene by "Click" Coupling. *J. Am. Chem. Soc.* **2005**, *127*, 14518–14524.

39 Rozkiewicz, D. I.; Jańczewski, D.; Verboom, W.; Ravoo, B. J.; Reinhoudt, D. N., "Click" Chemistry by Microcontact Printing. *Angew. Chem. Int. Ed.* **2006**, *45*, 5292–5296.

40 Agard, N. J.; Prescher, J. A.; Bertozzi, C. R., A Strain-Promoted [3 + 2] Azide–Alkyne Cycloaddition for Covalent Modification of Biomolecules in Living Systems. *J. Am. Chem. Soc.* **2004**, *126*, 15046–15047.

41 Dommerholt, J.; Rutjes, F. P. J. T.; van Delft, F. L., Strain-Promoted 1,3-Dipolar Cycloaddition of Cycloalkynes and Organic Azides. *Top. Curr. Chem.* **2016**, *374*, 16.

42 Song, W.; Wang, Y.; Yu, Z.; Vera, C. I. R.; Qu, J.; Lin, Q., A Metabolic Alkene Reporter for Spatiotemporally Controlled Imaging of Newly Synthesized Proteins in Mammalian Cells. *ACS Chem. Biol.* **2010**, *5*, 875–885.

43 Jin, Y.; Yu, C.; Denman, R. J.; Zhang, W., Recent Advances in Dynamic Covalent Chemistry. *Chem. Soc. Rev.* **2013**, *42*, 6634–6654.

44 Rowan, S. J.; Cantrill, S. J.; Cousins, G. R. L.; Sanders, J. K. M.; Stoddart, J. F., Dynamic Covalent Chemistry. *Angew. Chem. Int. Ed.* **2002**, *41*, 898–952.

45 Hu, J.; Gupta, S. K.; Ozdemir, J.; Beyzavi, H., Applications of Dynamic Covalent Chemistry Concept toward Tailored Covalent Organic Framework Nanomaterials: A Review. *ACS Appl. Nano Mater.* **2020**, *3*, 6239–6269.

46 Gutsche, C. D.; Iqbal, M., p-tert-Butylcalix[4] arene. *Org. Synth.* **1990**, *68*, 234.

47 Holler, M.; Allenbach, N.; Sonet, J.; Nierengarten, J.-F., The High Yielding Synthesis of Pillar[5] arenes under Friedel–Crafts Conditions Explained by Dynamic Covalent Bond Formation. *Chem. Commun.* **2012**, *48*, 2576–2578.

48 Feng, X.; Ding, X.; Jiang, D., Covalent Organic Frameworks. *Chem. Soc. Rev.* **2012**, *41*, 6010–6022.

49 Ding, S.-Y.; Wang, W., Covalent Organic Frameworks (COFS): From Design to Applications. *Chem. Soc. Rev.* **2013**, *42*, 548–568.

50 Waller, P. J.; Gándara, F.; Yaghi, O. M., Chemistry of Covalent Organic Frameworks. *Acc. Chem. Res.* **2015**, *48*, 3053–3063.

51 Gao, Q.; Li, X.; Ning, G.-H.; Xu, H.-S.; Liu, C.; Tian, B.; Tang, W.; Loh, K. P., Covalent Organic Framework with Frustrated Bonding Network for Enhanced Carbon Dioxide Storage. *Chem. Mater.* **2018**, *30*, 1762–1768.

52 Ulman, A., Formation and Structure of Self-Assembled Monolayers. *Chem. Rev.* **1996**, *96*, 1533–1554.

53 Chen, K.-Y.; Ivashenko, O.; Carroll, G. T.; Robertus, J.; Kistemaker, J. C. M.; London, G.; Browne, W. R.; Rudolf, P.; Feringa, B. L., Control of Surface Wettability Using Tripodal Light-Activated Molecular Motors. *J. Am. Chem. Soc.* **2014**, *136*, 3219–3224.

54 Arduengo, A. J.; Harlow, R. L.; Kline, M., A Stable Crystalline Carbene. *J. Am. Chem. Soc.* **1991**, *113*, 361–363.

55 Herrmann, W. A.; Köcher, C., N-Heterocyclic Carbenes. *Angew. Chem. Int. Ed. Engl.* **1997**, *36*, 2162–2187.

56 Dove, A. P.; Pratt, R. C.; Lohmeijer, B. G. G.; Li, H.; Hagberg, E. C.; Waymouth, R. M.; Hedrick, J. L., N-Heterocyclic Carbenes as Organic Catalysts. In *N-Heterocyclic Carbenes in Synthesis*, Nolan, S. P., Ed. Wiley-VCH: Verlag, **2006**; pp 275–296.

57 Crudden, C. M.; Horton, J. H.; Ebralidze, I. I.; Zenkina, O. V.; McLean, A. B.; Drevniok, B.; She, Z.; Kraatz, H.-B.; Mosey, N. J.; Seki, T.; Keske, E. C.; Leake, J. D.; Rousina-Webb, A.; Wu, G., Ultra Stable Self-Assembled Monolayers of N-heterocyclic Carbenes on Gold. *Nat. Chem.* **2014**, *6*, 409–414.

58 Inayeh, A.; Groome, R. R. K.; Singh, I.; Veinot, A. J.; de Lima, F. C.; Miwa, R. H.; Crudden, C. M.; McLean, A. B., Self-Assembly of N-heterocyclic Carbenes on Au(111). *Nat. Commun.* **2021**, *12*, 4034.

59 Dyke, C. A.; Tour, J. M., Solvent-Free Functionalization of Carbon Nanotubes. *J. Am. Chem. Soc.* **2003**, *125*, 1156–1157.

60 Chen, B.; Lu, M.; Flatt, A. K.; Maya, F.; Tour, J. M., Chemical Reactions in Monolayer Aromatic Films on Silicon Surfaces. *Chem. Mater.* **2008**, *20*, 61–64.

3

Molecular Recognition and Supramolecular Self-Assembly

3.1 History of Molecular Recognition and Supramolecular Chemistry

What is commonly referred to as **molecular recognition** is a process that occurs ubiquitously in the biological world. For example, a cell contains receptor proteins both on its surface and inside the cell which selectively bind to certain types of molecules, known as signaling molecules. The association or binding of a signaling molecule(s) with the receptor changes the receptor protein in some way, which in turn triggers a sequence of steps in which the signal (binding event) is amplified and transduced. Eventually, the signal transduction activates cellular responses to the signaling. The detection of signal molecules by receptor proteins relies on specific non-covalent interactions, including various non-covalent forces that have been discussed in Chapter 1. Therefore, molecular recognition is considered to be the specific interaction between two or more molecules through non-covalent forces, such as hydrogen bonding, metal coordination, electrostatic attraction, hydrophobic, van der Waals, and π–π interactions. Molecules that recognize one another in this way show molecular complementarity. Such a phenomenon is of great importance in the areas of molecular biology, crystal engineering, and supramolecular assembly.

Chemists have been aware of the phenomenon of molecular recognition for a long time. For example, Louis Pasteur in 1857 discovered that in incubations of ammonium ($\pm$)-tartrate with unidentified microorganisms, (+)-tartaric acid was consumed with considerable preference over (−)-tartaric acid. Pasteur's finding quickly sparked interest in the study of activity and stereoselectivity in biological and enzymatic reactions. In 1894 Emil Fischer proposed a "lock-and-key" metaphor to rationalize the enantioselectivity observed in enzyme-substrate interactions (Figure 3.1A). In Fischer's model, the substrate fits the binding site of an enzyme showing size and shape complementarity. This image represents a rigid

Organic Nanochemistry: From Fundamental Concepts to Experimental Practice,
First Edition. Yuming Zhao.
© 2024 John Wiley & Sons, Inc. Published 2024 by John Wiley & Sons, Inc.

lock-and-key model, in which the conformations of the substrate and enzyme are not supposed to undergo significant changes during the binding process. Later on, as greater understanding of various types of molecular recognition and binding phenomena was established, the lock-and-key model underwent numerous modifications. An important one is the **induced fit model**, where enzyme and substrate undergo significant conformational changes upon binding to one another. In 1933 Easson and Stedman formulated a **three-point-attachment** (TPA) hypothesis to rationalize enantioselectivity at adrenergic receptors, which has become widely accepted as the explanation for the selective binding of an enantiomer to a biological substrate (Figure 3.1B).

The observation of that manmade molecules can show molecular recognition was first made by Charles J. Pederson in 1967, two years before his retirement from DuPont. In synthesizing bis[2-(o-hydroxyphenoxy)ethyl] ether, Pederson used partially protected catechol that was contaminated with about 10% unprotected catechol. The reactions, as Pederson later described in his Nobel lecture, *gave a product mixture in the form of an unattractive goo. Initial attempts at purification gave a small quantity (about 0.4%) of white crystals which drew attention by their silky, fibrous structure and apparent insolubility in hydroxylic solvents* [1]. This minor product turned out to be the first crown ether discovered, dibenzo 18 crown-6, which was found to form stable complexes with alkali metal ions and ammonium ion. Pederson's work initiated the pursuit of molecular recognition using synthetic molecules, an area often referred to as the **host–guest chemistry**. Donald J. Cram substantially developed this concept by constructing a wide range of macrocyclic molecules, especially 3D molecular cages called **carcerands** or **hemicarcerands**, which have the capability of capturing other molecules inside their cavities. The pioneering work of Cram established the foundation of modern

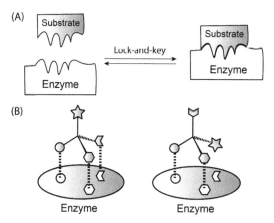

Figure 3.1 (A) Rigid lock-and-key model for enzyme-substrate binding. (B) Three-point-attachment for selective binding of an enantiomer to an enzyme.

host–guest chemistry. Jean-Marie Lehn, a French chemist well known for his study of 3D ligands called **cryptands**, introduced the term **supramolecular chemistry** in 1978 to generalize early developments in this field [2]. Since then, the essential concepts of supramolecular chemistry have been clearly established by extensive studies in order to understand the non-covalent interactions between a wide range of guest and host molecules. Pedersen, Cram, and Lehn together shared the Nobel Prize in Chemistry in 1987.

So, what is supramolecular chemistry? Lehn defined supramolecular chemistry as the *chemistry of molecular assemblies and of intermolecular bonds*. In other words, supramolecular chemistry is the chemistry beyond molecules. An organized complex entity resulting from the association of two or more chemical species held together by intermolecular forces is considered to be a supermolecule. Nevertheless, this expression is oversimplified, somewhat vague, and difficult to set boundaries for since any kind of system containing more than one molecule or ion would qualify as a supermolecule, and it could even be applied to a brine solution in a beaker. A more specific definition of supramolecular chemistry is described by the IUPAC golden book as follows, supramolecular chemistry is a field of chemistry related to species of greater complexity than molecules, that are held together and organized by means of intermolecular interactions. The objects of supramolecular chemistry are supermolecules and other polymolecular entities that result from the spontaneous association of a large number of components into a specific phase (membranes, vesicles, micelles, solid state structures etc.).

Supramolecular chemistry in general encompasses two areas, **host–guest chemistry** and **self-assembly**. The host–guest interaction refers to the scenario where an ion or small molecule (guest) is wrapped around or engulfed by another molecule (host) that is significantly larger in size. The binding sites of the host and guest molecules contain functional groups that enable complementary non-covalent interactions. Typically, the host molecule contains multiple ligands that are convergent in the complex, while the guest molecule possesses divergent partner groups which interact with the host. As illustrated in Figure 3.2, a host molecule can take a 1D linear, 2D macrocyclic, or a 3D cage-like shape. Numerous non-covalent forces may be involved in forming host–guest complexes, including metal coordination, hydrogen bonding, π-π stacking, and charge-transfer interactions. Moreover, the guest and host molecules must possess appropriate size, shape, and chemical nature to ensure effective complexation.

Self-assembly is a process where two or more molecular species interact through complementary non-covalent forces, forming organized clusters, aggregates, or crystalline structures eventually arriving at an equilibrium between the interacting species and the self-assembled product. The molecular building blocks used for self-assembly do not fit in the definition of "host" or "guest." Instead, they contain pairs of complementary groups non-covalently bound together. As such, the processes involved in self-assembly are usually highly spontaneous, with the

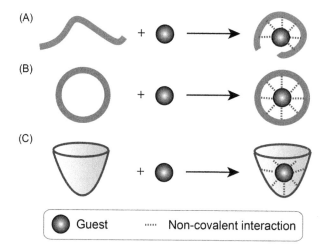

Figure 3.2 Formation of 1:1 host–guest complexes in which the host molecules take (A) acyclic chain, (B), macrocyclic, and (C) cage-like structures.

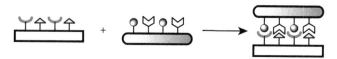

Figure 3.3 Schematic illustration of the self-assembly of two molecular building blocks through complementary non-covalent interactions.

outcomes being controlled by the structural information (size, shape, functionality) predesigned in the molecular building blocks (see illustrations in Figure 3.3).

Both host–guest chemistry and self-assembly play vitally important roles in modern nanochemistry. The following sections will outline the basic concepts and principles of molecular recognition and supramolecular self assembly.

3.2 Binding and Binding Constants

When a host–guest complex is formed in a solvent at a given temperature, its thermodynamic stability can be measured by the **binding constant**, K_a, associated with the complexation reaction. Strictly speaking, the binding constant should be calculated based on the activities (α) of the chemical species in equilibrium and therefore are dimensionless. In common practice, however, binding constants are often calculated approximately using concentrations and hence are reported with units. For example, the 1:1 complexation of a host (H) and a guest (G) in Eq. 3.1 gives a binding constant as calculated by Eq. 3.2, in which [H·G], [H], and [G] are the concentrations of host-guest complex, host, and guest, respectively.

The binding constant hence is expressed with a unit of M^{-1} or mol^{-1} L. It is also worth noting that both the solvent and the temperature have a significant impact on the binding constant. So, it is necessary to include these conditions when a binding constant is reported.

$$H + G \rightleftharpoons H \cdot G \tag{3.1}$$

$$K_a = [H \cdot G]/([H] \times [G]) \tag{3.2}$$

$$K_d = 1/K_a \tag{3.3}$$

$$\Delta G^\circ = -RT\ln(K_a) \tag{3.4}$$

$$\ln(K_a) = -\Delta H / RT + \Delta S / R \tag{3.5}$$

$$K_a = k_1 / k_{-1} \tag{3.6}$$

The binding constant can be called an **association constant** (K_a), a **stability constant** (K_s), or a **formation constant** (K_f). When the binding constant has been determined experimentally, the Gibbs energy (ΔG°) of the host–guest equilibrium reaction can be obtained using the Gibbs-Helmholtz equation (Eq. 3.4). The thermodynamic properties ΔH and ΔS are correlated to K_a and temperature (T) as shown in the van't Hoff equation (Eq. 3.5). Besides thermodynamic properties, K_a is related to kinetic rate constants as shown in Eq. 3.6, where k_1 and k_{-1} are the forward and reverse rate constants for the host–guest equilibrium reaction. Hence the binding constant can also be determined, if kinetic data are available.

In biological and pharmaceutical studies, another term, known as the **dissociation constant** (K_d), is more frequently used. K_d is simply the reciprocal of K_a, and it has a unit of M or $mol\ L^{-1}$ for a 1:1 host–guest complex. In this case, K_d equals the concentration where half of the complex (H·G) has dissociated, allowing the stability and binding affinity to be more intuitively assessed. For example, in the analysis of drug/receptor interactions, K_d is the concentration of the drug where 50% of the receptors are occupied. The smaller the K_d of a drug/receptor complex, the stronger the affinity the drug exhibits for the receptor.

Note that 1:1 complexation is not the only binding mode for host–guest interactions. Host–guest binding may result in a stable assembly containing m host molecules and n guest molecules, where at least one of the species has a coefficient (m or n) greater than one. In such cases, a binding constant β_{mn} is used to describe the overall process (see Eqs. 3.7 and 3.8).

$$mH + nG \rightleftharpoons H_m \cdot G_n \tag{3.7}$$

$$\beta_{mn} = [H_m \cdot G_n]/([H]^m \times [G]^n) \tag{3.8}$$

A common scenario is the stepwise 1:2 equilibria between one host molecule and two guest molecules, where the individual binding constants K_{a1} and K_{a2} are expressed as follows:

$$K_{a1} = [\text{H·G}]/([\text{H}]\times[\text{G}]) \tag{3.9}$$

$$K_{a2} = [\text{H·G}_2]/([\text{H}]\times[\text{H·G}]) \tag{3.10}$$

The overall binding constant, β_{12}, is therefore the product of the two individual constants, K_{a1} and K_{a2} (see Eq. 3.11). Typically, the magnitude of such binding constants is very large. Therefore they are often reported as the logarithm of β_{mn} (i.e., $\log \beta_{mn}$).

$$\beta_{12} = K_{a1} \times K_{a2} \tag{3.11}$$

Binding constants can be determined using a large variety of experimental methods (e.g., UV-Vis absorption, fluorescence emission, NMR, mass spectrometry, and calorimetry), as long as corresponding binding isotherms are established [3]. A **binding isotherm** is the theoretical change in the concentration of one component as a function of the concentration of another component at a constant temperature. Let's take 1:1 complexation as an example and given that the host and guest molecules have initial concentrations of $[\text{H}]_0$ and $[\text{G}]_0$ after mixing, respectively. Their relationships to the concentration of the complex $[\text{H·G}]$ can be expressed as follows:

$$[\text{H}]_0 = [\text{H}]+[\text{H·G}] \tag{3.12}$$

$$[\text{G}]_0 = [\text{G}]+[\text{H·G}] \tag{3.13}$$

where [H] and [G] are the equilibrium concentrations of free host and guest molecules. The binding constant, K_a, can be then derived from the above two equations as expressed in Eq. 3.14.

$$K_a = [\text{H·G}]/\{([\text{H}]_0 - [\text{H·G}])\times([\text{G}]_0 - [\text{H·G}])\} \tag{3.14}$$

Based on the equation above, binding isotherms for varying K_a values can be plotted as illustrated in Figure 3.4. It is worth noting that even for the simplest 1:1 binding mode, the correlation between the binding constant and concentrations of [H·G] follows a complex nonlinear relationship. As such, binding systems with different binding constants show isotherm curves with very different shapes. The shape of an isotherm curve is dependent on both the binding constant and the initial concentration of the host or guest molecules. When the binding constant is very large and the initial concentration is high, the isotherm plot would not show significant curvature and therefore the binding constant cannot be meaningfully

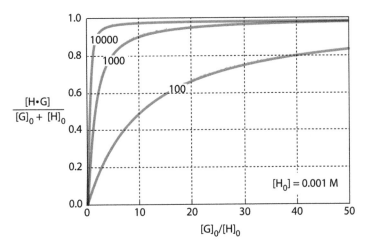

Figure 3.4 Theoretically simulated binding isotherms for varying K_a values (100 M^{-1}, 1000 M^{-1}, and 10,000 M^{-1}) with [H]$_0$ = 0.001 M.

measured under such conditions. As a rule of thumb, the concentration of the host molecule should match its dissociation constant, K_d, in order to have the binding isotherm show a curvature for reliable binding constant analysis.

Determination of a binding constant(s) can be carried out by fitting an experimentally established binding isotherm with a certain equilibrium model (e.g., 1:1, 1:2, or 2:1 binding) with the aid of a range of nonlinear fitting software packages, such as *SPECFIT* and *Bindfit*. It is also worth mentioning that there exist various methods, in the classical literature, which transform the nonlinear relationships into linear regressions. Among them, the Benesi-Hildebrand [4] and Scatchard [5] plots are well-known methods. These linear regression methods allow binding constants to be calculated based on different assumptions or approximations in the conditions of the experimental measurements, but they are often associated with problems of relatively large errors or distortion of the results. In modern research work, finding the exact solution of the nonlinear equation for a binding isotherm has become routine and has been widely adopted in lieu of the old-fashioned linear transformation approaches.

In experimental studies of binding properties, the binding stoichiometry needs to be determined first so that the correct equilibrium model can be subsequently applied to extract reasonable binding constants from experimental binding isotherms. There are several ways to determine the stoichiometry, such as the continuous variation, slope ratio, and mole ratio methods. Among them, the continuous variation method (also known as the **Job's plot** method) is the most popular one. In this method, the sum of [H]$_0$ and [G]$_0$ is kept as a constant. A series of samples of host–guest complexes are prepared, in which

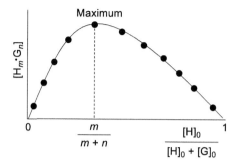

Figure 3.5 Illustration of an exemplar Job's plot.

the amounts of host and guest compounds are varied in such a way that the ratio, $[H]_0/([H]_0 + [G]_0)$ or $[G]_0/([H]_0 + [G]_0)$, covers the range of 0–1 as much as possible. Next, the concentration of the complex $(H_m \cdot G_n)$ in each sample needs to be determined through experimental methods such as UV-Vis, fluorescence, and NMR. With these data, a correlation curve is generated by plotting the concentration of $H_m \cdot G_n$ against the mole fraction, $[H]_0/([H]_0 + [G]_0)$ or $[G]_0/([H]_0 + [G]_0)$. This plot is popularly known as the Job's plot (see Figure 3.5). When the x-axis of the plot is $[H]_0/([H]_0 + [G]_0)$, the maximum of the curve corresponds to ratio of $m/(m + n)$ for the complex $H_m \cdot G_n$. It is therefore predicted that when 1:1 complexation occurs, the maximum of the Job's plot is at 0.5. For 1:2 and 2:3 binding modes, the maxima should be observed at 0.33 and 0.40, respectively. In the experimental work, the y-axis of the Job's plot does not always have to be accurately determined concentration of the complex $H_m \cdot G_n$. As long as an experimentally observable property (e.g., UV-Vis absorbance, NMR chemical shift or integral) shows a linearly proportional relationship to the concentration of $H_m \cdot G_n$, it can be used to generate the Job's plot.

As discussed above, the concentration of the host–guest complex must be accurately determined in order to evaluate the binding stoichiometry and binding constants. Commonly used experimental methods are UV-Vis absorption and NMR titration analyses. For example, UV-Vis absorption analysis is a very sensitive technique used to detect various chemical species in the solution phase. Under dilute conditions (e.g., 10^{-5} to 10^{-4} M), a colored species would obey the Beer's Law, $A = \varepsilon \times l \times C$, in which its experimentally measured absorbance (A) is linearly proportional to its concentration (C). In the Beer's law equation, ε is the molar absorptivity (or extinction coefficient) of the compound at a certain wavelength, and l is the path length of the sample cuvette. Given that a host (H), a guest (G), and their complex (E) show different UV-Vis absorption profiles as illustrated in Figure 3.6, the absorbance at the maximum absorption wavelength in the observed spectrum (A_{obs}) can be determined by means of Eqs. 3.15–3.19. Note that

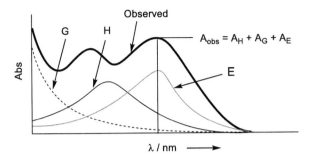

Figure 3.6 UV-Vis absorption spectra of a host (H), a guest (G), and their complex (E) in correlation to the observed spectrum of their mixture.

the path length l is assumed to be 1 cm and hence l is "left out" of the Beer's law equations below.

$$A_H = \varepsilon_H \times [H] = \varepsilon_H \times \left([H]_0 - m[E]\right) \tag{3.15}$$

$$A_G = \varepsilon_G \times [G] = \varepsilon_G \times \left([G]_0 - n[E]\right) \tag{3.16}$$

$$A_E = \varepsilon_E \times [E] \tag{3.17}$$

$$A_{obs} = A_H + A_G + A_E$$

$$A_{obs} = \varepsilon_H \times \left([H]_0 - m[E]\right) + \varepsilon_G \times \left([G]_0 - n[E]\right) + \varepsilon_E \times [E] \tag{3.18}$$

In these equations, A_H, A_G, and A_E are the absorbances of host, guest, and complex, respectively; ε_H, ε_G, and ε_E are the molar absorptivities of host, guest, and complex, respectively. A_{obs} is the absorbance experimentally observed. Eq. 3.18 can be further transformed into Eq. 3.19 to show a term, $A_{obs} - \varepsilon_H \times [H]_0 - \varepsilon_G \times [G]_0$, which is linearly proportional to the concentration of complex. Herein, A_{obs}, ε_H, ε_G, $[H]_0$, and $[G]_0$ can all be experimentally determined. Therefore, this term can be applied to the generation of the Job's plot as well as in the nonlinear fitting analysis of the experimental binding isotherm.

$$\{A_{obs} - \varepsilon_H \times [H]_0 - \varepsilon_G \times [G]_0\} = (\varepsilon_E - m \times \varepsilon_H - n \times \varepsilon_G) \times [E] \tag{3.19}$$

NMR spectroscopic analysis provides another convenient and reliable experimental method for investigating host–guest binding behavior. For example, ^1H and ^{19}F NMR titration experiments can provide useful information about the

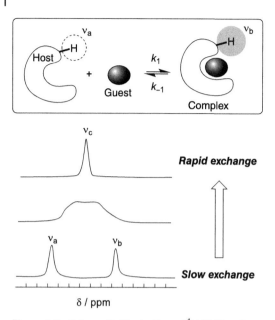

Figure 3.7 Schematic illustration of ^{1}H NMR patterns corresponding to a 1:1 host–guest complexation process at varying exchange rates.

binding mode and binding isotherm. Variable temperature (VT) NMR analysis provides insight into the kinetic properties involved in host–guest binding and other supramolecular processes. Nuclear Overhauser effect (NOE) and diffusion-ordered (DOSY) NMR experiments allow through-space interactions and diffusion coefficients to be investigated.

Unlike UV-Vis absorption analysis, NMR experiments feature a relatively long time scale and therefore determine an exchange process in a time-dependent manner. Let's take a 1:1 host–guest binding process as an example (see Figure 3.7). Assume a proton in the free host molecule gives a resonance frequency at ν_a. When binding takes place, the chemical environment of this proton is altered. The resonance frequency ν_a is therefore shifted to a different value ν_b. If both the complex and host molecules have long lifetimes, in other words, the exchange between the 1:1 complex and free host and guest molecules is a slow process, the ^{1}H NMR spectrum for a sample in which both free host and complex are present would clearly show two distinct signals at ν_a and ν_b, respectively. In this case, the relative concentrations of the complex and the host can be simply determined from the integral ratios corresponding to these two signals. With these data, the binding constant for the host–guest interaction can be established.

$$\Delta\upsilon = (h/2\pi) \times \tau \tag{3.20}$$

$$\upsilon_c = \left[a/(a+b)\right] \times \upsilon_a + \left[b/(a+b)\right] \times \upsilon_b \tag{3.21}$$

Most binding events, however, are not slow exchange scenarios. According to the Heisenberg uncertainty principle (Eq. 3.20), the uncertainty in determining an NMR resonance frequency (Δv) is proportional to the lifetime (τ) of the species measured by NMR. When the exchange rate matches the ^{1}H NMR time scale τ (usually 0.4 ms to 0.2 s), the uncertainty in determining the resonance frequency (Δv) becomes so significant that signal coalescence occurs as a result of line broadening of the two signals involved in the exchange. When the exchange rate is faster than the ^{1}H NMR time scale, the two different protons in exchange cannot be differentiated by NMR analysis, even though they are in different chemical environments. Instead, the two protons would merge as one signal, the resonance frequency (or chemical shift) of which is a mole-fraction weighted average of the two resonances as described in Eq. 3.21. In this equation, v_a and v_b are the resonance frequencies of the host and complex, respectively, v_c is experimentally observed frequency of the host–guest mixture, a and b are the populations of free host and host–guest complex, respectively. In the fast exchange scenario, the populations of host and complex molecules can still be calculated from ^{1}H NMR analysis, if the resonance frequencies of the free host and complex molecules are known.

Figure 3.8 illustrates the results of ^{1}H NMR titration experiments where a cage-like cation host, namely a cryptand, is complexed with sodium cation at different host–guest molar ratios. As can be seen from Figure 3.8B that the chemical shifts

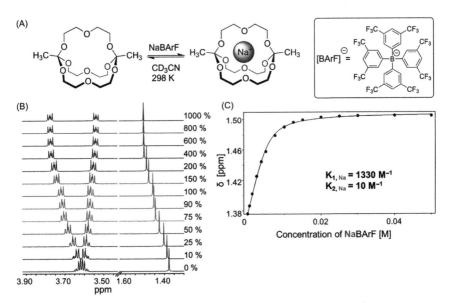

Figure 3.8 (A) Complexation of sodium cation with a cryptand. (B) Partial ^{1}H NMR spectra (400 MHz, CD$_3$CN, 298 K) monitoring the titration of the cryptand (5 mM) with sodium cation from 0 to 1000 mol%. (C) Binding isotherm of sodium ion-cryptand complexation. Binding constants K_1 and K_2 were obtained from fitting in 1:1 and 1:2 models, respectively. (Reproduced from Int. J. Mol. Sci. 2015, 16, 20641-20656 / MDPI / CC BY 4.0).

of the ^{1}H signals of the host gradually change in response to the sodium cation titration [6]. Figure 3.8C shows the binding isotherm based on monitoring the chemical shift of the methyl protons in the cryptand. Fitting the isotherm with two complexation models, 1:1 and 1:2, give two binding constants, $K_1 = 1330$ M^{-1} and $K_2 = 10$ M^{-1}, respectively.

Besides spectroscopic analyses, detection of the heat change during a non-covalent binding process can also provide information about the binding stoichiometry and binding constant. One widely used method is called **isothermal titration calorimetry** (ITC), which is nowadays regarded as the gold-standard technique for studying intermolecular interactions [7]. The working principle of an ICT analysis is based on the measurement of the heat generated or absorbed resulting from the binding between two molecules (i.e., an exothermic or an endothermic process). An ITC instrument is schematically illustrated in Figure 3.9. The instrument has two cells, one of which contains the host compound dissolved in a solvent, and the other contains only the solvent to act as a reference. Both cells are kept at constant temperature and pressure during the measurement. The titration is carried out by repetitively injecting a solution of guest compound into the sample cell. The binding of host and guest compounds causes heat release or absorption, which is monitored by the sensing devices of the instrument. The instrument compensates for heat lost or gained as a result of the binding process maintaining the temperature of the sample cell the same as the reference cell. In this way, the heat flow associated with the binding process is determined.

Saturation of host–guest binding is reached when there is no further heat discharged or consumed. Once the titration is finished, the heat flow detected is calculated by integrating the power over the time to give the binding enthalpy (ΔH_b). The heat discharged or consumed as the binding process proceeds corresponds to

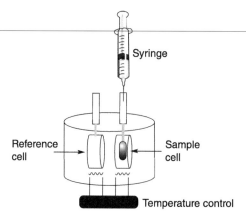

Figure 3.9 Schematic illustration of an ITC instrument.

the increase in host–guest concentration. The total heat (Q) released or absorbed is expressed by the following equation:

$$Q = V\Delta H_b \left[H \cdot G \right] \tag{3.22}$$

where V is the volume of the cell, ΔH_b is the enthalpy of binding, and $[H \cdot G]$ is the concentration of the resulting host–guest complex. Using the 1:1 binding model, Eq. 3.22 can be transformed into the following equation,

$$Q = V\Delta H_b K_a \left[H \right]_0 \left[G \right] / \left(1 + K_a \left[G \right] \right) \tag{3.23}$$

where K_a is the 1:1 binding constant, $[H]_0$ is the initial concentration of the host, and $[G]$ is the concentration of the uncomplexed guest molecules. With these equations, the ITC binding isotherm can be established. Nowadays, as a result of extensive development over the years, the ITC technique exhibits very high sensitivity, making it highly suitable for various research tasks, ranging from protein–ligand binding, metal–ligand interactions, DNA/DNA and protein/DNA interactions, to polymer surfactant interactions, and just mention a few. Compared with UV-Vis absorption and NMR analyses, ITC has the advantage of being able to determine both the binding constant and the binding enthalpy from a single experiment. The stoichiometry for the binding reaction can also be accurately established from an ITC analysis under certain conditions. Conducting ITC experiments at different temperatures allows the change in heat capacity to be determined, so that a full thermodynamic characterization of the binding process can be obtained.

Figure 3.10 shows the experimental ITC results for the 1:1 binding of a compound named buckycatcher and [60]fullerene (C_{60}) [8]. The molecular structure of the buckycatcher contains two bowl-shaped corannulene units tethered through a tet-rabenzocyclooctatetraene unit. As such, the molecule is preorganized for host–guest interactions with C_{60} (buckyball) through concave–convex complementarity. The ICT raw data was fitted with a 1:1 complexation model, eliciting a binding constant, $K_a = 3200 \pm 150$ M^{-1}, while related thermodynamic data for the binding were also determined as $\Delta G = -4.77 \pm 0.03$ kcal mol^{-1}, $\Delta H = -4.61 \pm 0.1$ kcal mol^{-1}, and $-T\Delta S = -0.16 \pm 0.12$ kcal mol^{-1}.

3.3 Cooperativity and Multivalency

When describing how supramolecular chemistry works, an analogy of a sports team can be made; that is, players on the team through collaboration achieve an overall performance that is much greater than the sum of individual players. Just as the ancient Greek philosopher Aristotle said, "the whole is greater than the sum of its parts". A supramolecular system is created and/or organized by means of the

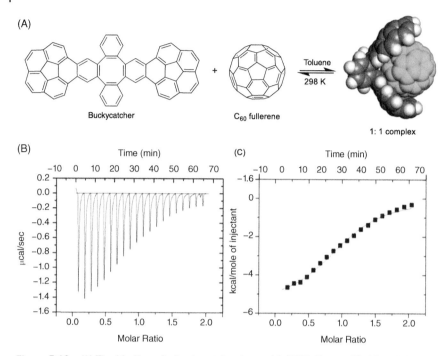

Figure 3.10 (A) The binding of a buckycatcher host with [60]fullerene (C_{60}) in a 1:1 binding mode. (B) ITC raw data for the titration of the buckycatcher with C_{60}. The buckycatcher solution (0.7 mM in toluene) was titrated into the ITC cell filled with C_{60} (70 μM in toluene) through 20 separate injections. (C) Nonlinear regression fitting of ΔH for each injection with the 1:1 complexation model. (Reproduced from *J. Phys. Chem. B* 2014, *118*, 11956–11964).

association of two or more molecular building blocks held together by various types of intermolecular forces, such as hydrogen bonding, π–π stacking, hydrophobic, and coordination interactions. These interactions are not only additive but also cooperative. As a result, the properties of a supramolecular assembly are different than (often outperform) the sum of the properties of each individual component.

Many biological systems such as enzymes, receptors, and ribosomes show cooperative binding properties, known as the **allosteric interactions** or **allostery**. Allostery refers to the modulation of the binding at an active site as a consequence of the binding at a remote (effector) site. When the active and effector sites are equivalent, the allostery is called **cooperativity** or **homotropic allostery**. The binding of hemoglobin with oxygen is a paradigm example of **cooperativity**, which has generated enormous scientific interest since 1904. Christian Bohr, a Danish physician and the father of the 1922 Nobel Prize laureate in physics, Niels Bohr, was the first to notice the sigmoidal nature of oxygen–hemoglobin binding, now called the **Bohr effect**. C. Bohr's observation led to the discovery that hemoglobin can undergo conformational changes to increase its affinity for oxygen as oxygen molecules progressively bind to each of its four available binding sites, showing a propensity for **positive cooperativity**.

In supramolecular chemistry, cooperativity is an important concept that describes how the binding of one ligand influences the affinity of a receptor in its further binding interactions with other ligands [9]. Binding site cooperativity in host–guest interactions is a generalization of the **chelate effect**, a well-known concept in classical coordination chemistry [10]. The chelate effect is simply illustrated by the examples shown in Figure 3.11, in which the Ni(II) complex with three bidentate ethylenediamine (en) ligands is 10^8 times more stable than that the Ni(II) complex containing six monodentate ammonia ligands.

The chelate effect arises from a number of factors, among which the enthalpic and entropic contributions play very significant roles. On one hand, the double-ended (bidentate) ligand has two binding interactions in comparison to the single interaction effected by a simple monodentate ligand. So, the Gibbs energy (ΔG) of bidentate binding is much stronger than monodentate binding. On the other hand, the bidentate binding gains entropy advantages over the monodentate binding. Simply put, when one end of a bidentate ligand is coordinated with the metal, the other ligand end is already in the vicinity and can readily bind without paying as much the translational and rotational entropy cost as the first binding interaction. The entropic contribution on the chelation effect can also be understood by the ligand exchange reaction shown in Figure 3.12.

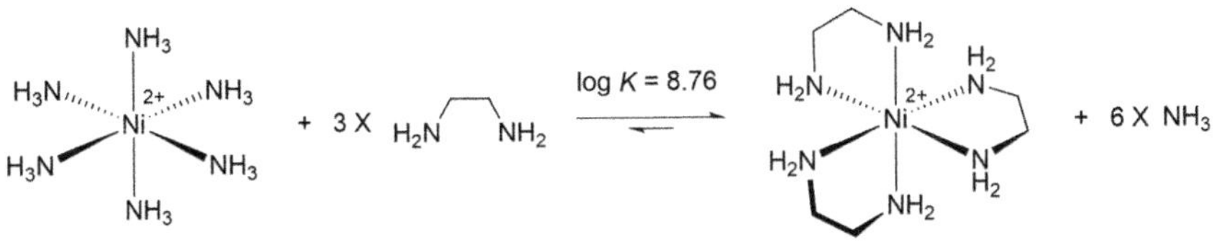

Figure 3.11 Equilibrium between $Ni(NH_3)_6^{2+}$ and $Ni(en)_3^{2+}$ as a result of the chelate effect.

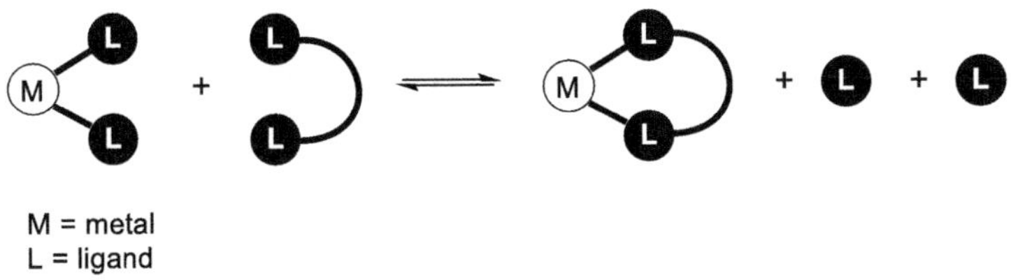

M = metal
L = ligand

Figure 3.12 Exchange reaction of two monodentate ligands and a tethered bidentate ligand.

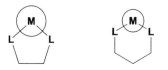

Larger bite size   Smaller bite size

Figure 3.13 Comparison of the geometries of five- and six-membered chelate rings.

$$M + L \xrightarrow{K_1} ML$$

$$ML + L \xrightarrow{K_2} ML_2$$

$$\vdots$$

$$ML_{i-1} + L \xrightarrow{K_i} ML_i$$

$$ML_{n-1} + L \xrightarrow{K_n} ML_n$$

Figure 3.14 Multistep binding of a metal cation (M) with n ligands (L).

Assuming that all the ligand–metal bonds (M–L) and the solvation energies of the complexes and ligands are equivalent, the enthalpy change ($\Delta H°$) of this model reaction is zero. The formation of chelate product results in the release of two moles of free monodentate ligands, hence causing the entropy change ($\Delta S°$) to be positive and the Gibbs energy change ($\Delta G°$) to be significantly negative. It has been estimated that one mole of solute at the standard state concentration of 1.00 molal gives a $\Delta S°$ of 7.9 e.u. (1 entropy unit e.u. = 4.184 J K^{-1} mol^{-1}), contributing about two log K units to the equilibrium constant at 25 °C.

The stabilization caused by the chelate effect is highly dependent on the size of the chelate ring. As shown in Figure 3.13, a five-membered ring is usually the most favored size, as it affords not only negligible strain energy but the optimum "bite size" for coordination with cations, particularly large metal cations. A six-membered ring is also relatively free of strain, but the distance between the two tethered ligands is shorter than that of five-membered ring. As a result, six-membered-ring chelation is favored for smaller metal ions. Chelate complexes with other ring sizes are not observed as often as five- and six-membered rings. For example, a four-membered-ring chelate structure is highly strained and therefore would not afford stable complexes.

In host–guest chemistry, chelate binding or cooperative binding is frequently observed. Cooperativity can be examined using the stepwise binding of a metal cation with multiple ligands shown in Figure 3.14.

Eq. 3.24 serves as a measure for evaluating the cooperativity of the system.

$$K_{i+1} / K_i = i(n-i) / \left[(i+1)(n-i+1) \right] \tag{3.24}$$

In this equation, K_{i+1} and K_i denote the binding constants of the $(i+1)$th and ith steps. If K_{i+1}/K_i is higher than calculated from Eq. 3.24, the system exhibits positive (synergetic) cooperativity, meaning that the subsequent binding is stronger than that of the previous one. If K_{i+1}/K_i is lower than expected from Eq. 3.24, the system exhibits negative cooperativity (interfering). If K_{i+1}/K_i is the same as calculated from Eq. 3.24, the system shows a noncooperative (additive) effect.

If a host (or receptor) and a guest (or ligand) both contain multiple binding sites that are connected through spacers, they are called **multivalent** and their interaction is hence termed **multivalent binding. Multivalency** is a key principle in supramolecular and biological chemistry for achieving strong, but reversible interactions [11]. The hook-and-loop fastener, Velcro®, is an excellent analogy from daily life for multivalent binding; that is, a single hook does not shoulder much of a load and is easily released, but many together form a connection resisting even strong separation forces. In the field of biochemistry, another term, **avidity**, is often used, which shares the same meaning as multivalency. The concept of avidity was first introduced to describe the binding behavior of immunoglobins with multiple valences, and now is commonly used to refer to the accumulated strength of multiple affinities between a protein receptor and its ligand, such as the binding of an antibody with a complex antigen. The terms of multivalency and avidity are synonymous; however, multivalency is more commonly used in the field of organic and supramolecular chemistry.

Whitesides proposed an enhancement factor β to characterize a multivalent binding effect, which can be expressed by the following equation,

$$\beta = K_{multi} / K_{mono} \tag{3.25}$$

where K_{multi} is the binding constant for the binding of a multivalent ligand to a multivalent receptor, K_{mono} is the binding constant for the binding of a monovalent ligand to a multivalent receptor. An advantage of this enhancement factor is that it can be used even if the multiplicity of effective binding interactions is unknown. A disadvantage is that it simultaneously includes the influence of the cooperativity and the symmetry effect. It is worth mentioning that multivalency and cooperativity are different concepts. Multivalent interactions do not always have to be associated with positive cooperativity. A multivalent binding may still be useful even if the binding shows negative cooperativity.

For synthetic multivalent hosts and guests, the number of binding sites can be controlled during synthesis or determined in a final stage using various analytical methods as previously discussed. Hence individual contributions to the multivalent interaction can be determined. The analysis of **effective molarity** (*EM*) is quite informative in this respect [12]. *EM* in general is used to describe the intramolecular reactivity corrected for the inherent reactivity. The concept of *EM* can be explained by an example of the divalent interaction shown in Figure 3.15.

Assuming the binding constant for a monovalent receptor and a monovalent ligand is K_{momo} (Figure 3.15A), the binding of a divalent receptor and a monovalent ligand should ideally give a binding constant $K_1 = 2K_{mono}$, since each divalent receptor molecule is equivalent to two monovalent receptor molecules (Figure 3.15B). The subsequent binding of another ligand then leads to a 1:2 complex with a binding constant $K_2 = 1/2K_{mono}$. The cumulative constant K is therefore calculated as $(K_1 \times K_2) = (K_{mono})^2$. In the case of a divalent receptor binding to a divalent ligand in a stepwise manner (Figure 3.15C), the binding constant for the first

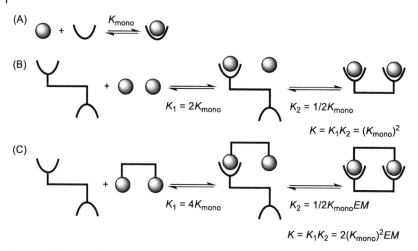

Figure 3.15 Equilibria for the binding of (A) a monovalent receptor and a monovalent ligand, (B) a divalent receptor and two monovalent ligands, and (C) a divalent receptor and a divalent ligand. K_{mono}, K_1, and K_2 are individual binding constants. K is the total binding constant.

step is $4K_{mono}$. The second binding step outlined in this scheme is an intramolecular binding, forming a cyclic complex. The binding constant for the second step is expressed as $(1/2K_{mono} \times EM)$, in which EM is a parameter governing the preference for an intermolecular or an intramolecular complexation process. If the concentration of the substrate (e.g., receptor) is higher than EM, the intramolecular process cannot effectively compete with the intermolecular process. Moreover, the term $(K_{mono} \times EM)$ is an expression of cooperativity. When this term is greater than one, the divalent binding shows **positive cooperativity**. When this term is less than one, the divalent binding shows **negative cooperativity**, and the resulting complex is mostly formed through the intermolecular approach.

3.4 Preorganization and Complementarity

Many host–guest complexes achieve remarkably higher stabilities than would be expected from the chelate or cooperativity effect. This effect has been widely found among the macrocyclic and macrobicyclic hosts, traditionally known as the **macrocyclic effect**. A classic example used to illustrate this effect is the pair of Cu(II) complexes reported by Cabbiness and Margerum in 1969 (Figure 3.16) [13]. Although both complexes contain the same type and number of N–Cu(II) interactions, the macrocyclic ligand on the left affords a complex about 10^4 times more stable than its acyclic analogue does. This extra stabilization delivered by the macrocyclic ligand can be attributed to the organization of the binding sites in space prior to binding with the guest, which is known as **preorganization** [14]. The structure of the acyclic ligand is flexible and differs significantly from the one in complexation with Cu(II) ion. To have the acyclic ligand wrapped around the metal ion for binding,

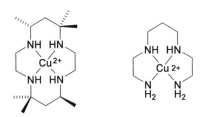

Figure 3.16 Cu(II) complexes with a macrocyclic ligand (left) and an analogous acyclic ligand (right).

a portion of energy cost needs to be paid for restricting rotation and translation as well as bringing the N sites to proximity (repulsive interactions). For the macrocyclic ligand, such an energy penalty has already been paid during its synthesis. The macrocyclic complex is thus thermodynamically more stable than the other one.

It can be simply stated that a host molecule (e.g., a macrocycle or a macrobicycle) that does not undergo a significant conformational change as a result of binding a guest molecule is **preorganized**. The higher the degree of preorganization a host exhibits, the greater the stability gained by the host–guest complex formed. Another good example of highly preorganized macrocyclic ligand is the crown ether. As shown in Figure 3.17, the complexation of a macrocyclic ligand, [18]crown-6, with various metal ions gives binding constants a few log K units larger than those for acyclic pentaglyme. [18]Crown-6 is a macrocycle with multiple polar C–O bonds. To minimize dipole–dipole repulsions, it prefers to adopt a somewhat folded conformation in the solid state (Figure 3.18A) or in low dielectric constant solvents. If dissolved in a strongly polar solvent, such as water, the solvation effect would lessen the dipole–dipole interactions. Either way, free [18]crown-16 needs to open up to host a metal ion in a D_{3d}-like conformation as shown in its 1:1 complex with potassium cation (Figure 3.18B). Still,

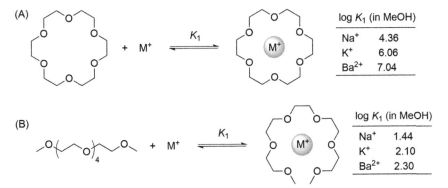

Figure 3.17 Complexation of (A) [18]crown-6 and (B) pentaglyme with various metal cations and related binding constants determined in methanol.

(A)   (B)

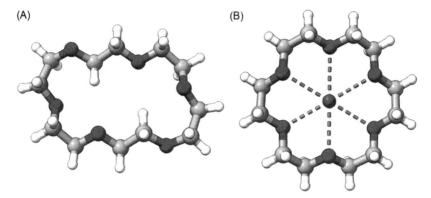

Figure 3.18 Molecular structures of (A) [18]crown-6 and (B) 1:1 complex of [18]crown-6 and potassium cation determined by X-ray single crystallography.

when compared with acyclic pentaglyme, [18]crown-6 is more preorganized, making it show a much stronger binding affinity for metal cations.

It is known that the oxygen lone pairs of [18]crown-6 are weakly solvated, particularly in protic solvents. So, in order to form a complex with a metal cation, [18]crown-6 needs to undergo a desolvation process in addition to a certain degree of conformational reorganization. These changes impact its binding affinity for metal cations. Cram and co-workers in 1981 developed a macrocyclic ligand, called **spherand**, which is fully preorganized for selective binding of smaller alkali metal cations like Li^+ and Na^+ [15]. As shown in Figure 3.19, the spherand was synthesized from a terphenyl precursor through a sequence of steps. Spherand features a rigid macrocyclic structure as illustrated by the X-ray crystal structure in Figure 3.20A. The six ethereal oxygen atoms are organized in an octahedral arrangement, pointing toward the center of the macrocyclic cavity, and are hence prevented from being solvated. Upon complexation with lithium cation, the spherand barely shows any conformational changes (Figure 3.20B). The full preorganization of spherand-6 enables it to exhibit an extremely large binding affinity for a suitable cation host; for example, the binding constant (K_a) for Li^+ ion is greater than 7×10^{16} M^{-1} and for Na^+ is 1.2×10^{14} M^{-1} measured in $CDCl_3$ (saturated with D_2O) at 25 °C. The binding with Li^+ is actually too strong to be precisely determined by the NMR titration method. It has been estimated that the Li^+/Na^+ selectivity is greater than 600 in favor of Li^+. On the other hand, spherand-6 does not show binding with larger cations, such as K^+, Rb^+, Cs^+, and NH_4^+, according to experimental observations.

Cryptands are variants of crown ethers showing a 3D molecular structure that can fully encapsulate or entomb a cation guest within their cavity. Cryptands were initially designed and studied by Lehn and co-workers, with the purpose of achieving stronger binding and greater selectivity for metal cations. Cryptands exhibit

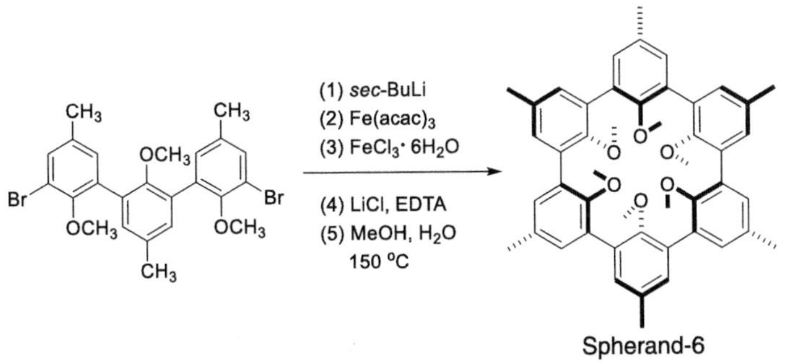

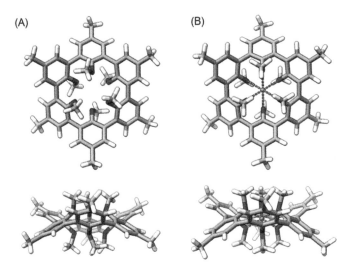

Figure 3.19 Cram's synthesis of spherand-6.

Figure 3.20 Molecular structures of (A) spherand-6 and (B) 1:1 complex of spherand-6 and lithium cation as determined by X-ray single crystallography.

the **cryptate effect** that facilitates their complexation with metal cation guests. Figure 3.21 shows an exemplar [2.2.2]cryptand and its two monocyclic analogues. The presence of an additional bridge in the 3D structure of the cryptand greatly reduces the freedom of motion of the ligand groups and lowers the degree of solvation of the donor atoms due to their proximity to one another. These factors work together to better stabilize the complex of the cryptand with a metal cation guest (e.g., K^+) in comparison to its monocyclic diazacrown ether analogues.

In addition to preorganization, other factors also play critical roles in determining the stability of a host–guest complex. One of the important factors is called

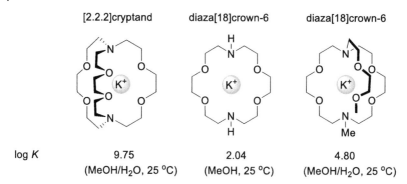

[2.2.2]cryptand	diaza[18]crown-6	diaza[18]crown-6
log K		
9.75	2.04	4.80
(MeOH/H$_2$O, 25 °C)	(MeOH, 25 °C)	(MeOH/H$_2$O, 25 °C)

Figure 3.21 Comparison of the binding affinities for potassium ion between a [2.2.2] cryptand and two analogous diazacrown ethers.

complementarity. Cram described the principle of complementarity as "to complex, hosts must have binding sites which can simultaneously contact and attract the binding sites of the guests without generating internal strains or strong nonbonded repulsions" [16]. Generally speaking, a complementary host–guest binding is achieved by having the host carry binding sites that are of the correct geometric and electronic characteristics (e.g., polarity, hydrogen bonding, hardness/softness) that match those of the guest. The lock-and-key model shown in Figure 3.1A is an example of shape and size complementarity between the active pocket of an enzyme and a substrate. Hydrogen bonded supramolecular assemblies are well known to show complementary hydrogen-bonding interactions. Figure 3.22 illustrates a group of DDD–AAA hydrogen-bonding pairs. As discussed in Chapter 1 (Section 1.4.2), the DDD–AAA type of interactions avoids disfavored secondary hydrogen-bonding interactions. As a result, these triply hydrogen-bonding pairs are associated with exceedingly large binding constants.

Complementarity can also be found in systems where the spatial arrangements of host and guest molecules act as a determining factor. Figure 3.23 shows an enantiomerically selective cyclophane host developed by Cram and co-workers in 1977 [17]. The cyclophane was synthesized from binaphthol precursors to show (S,S) chirality. When binding with a pair of chiral ammonium salts, this cyclophane host shows a higher affinity for the (R)-enantiomer than the (S)-enantiomer. The extent of chiral recognition exhibited by this cyclophane host is 0.266 kcal mol^{-1}, which is modest. However, the concept of spatial complementarity demonstrated by this example has led to the development of many useful chiral resolution and chiral catalytic methodologies.

Preorganization and complementarity are two principal effects that drive the formation of various host–guest complexes and supramolecular assemblies. Preorganization of the binding sites in the host molecule facilitates its binding with suitable guest molecules or ions. Complementarity, on the other hand, is

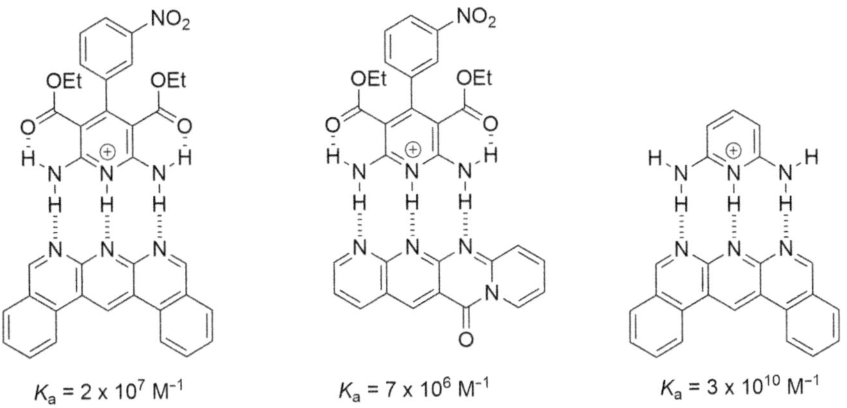

$K_a = 2 \times 10^7$ M^{-1} $K_a = 7 \times 10^6$ M^{-1} $K_a = 3 \times 10^{10}$ M^{-1}

Figure 3.22 Triply hydrogen-bonded pairs showing complementary DDD–AAA type of interactions.

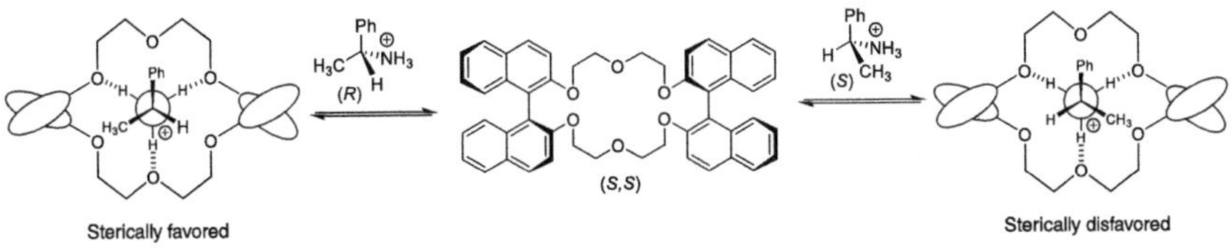

Sterically favored   Sterically disfavored

Figure 3.23 An enantiomerically pure cyclophane host that shows chiral recognition for ammonium salts.

necessary but not sufficient to ensure strong interactions between the host and guest species. Solvation also exerts a significant effect on the preorganization. All these factors operate in concert to control the binding selectivity and affinity for host–guest systems. These principles have been popularly used nowadays in the design of chemo-/biosensors, organic catalysis, and advanced molecular devices. It is also worth mentioning that in the context of dynamic covalent chemistry, a concept called **predisposition** was introduced by Sanders and co-workers in 1997 [18]. Predisposition should not be confused with the term *preorganization*. Predisposition can be viewed as a strong conformational and structural preference expressed by the building blocks once they are incorporated into a larger structure. An example of predisposition is illustrated in Figure 3.24, where a mixture of cinchonidine hydroxyester and xanthene hydroxyester underwent transesterification to exclusively form a homotrimeric and a homodimeric macrocyclic product. No heteromeric macrocyclic products were observed in this reaction, indicating the strong predisposition of the two starting compounds to form homomeric macrocyclic products in such thermodynamically controlled macrocyclization reactions.

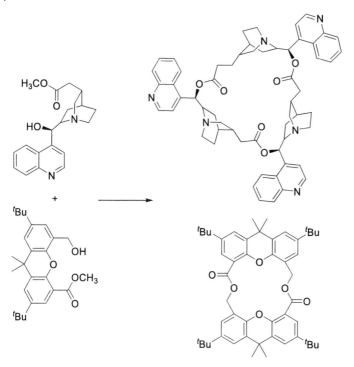

Figure 3.24 Thermodynamically controlled transesterification reactions of a mixture hydroxyester compounds showing predisposition for homomeric macrocyclic products.

3.5 Thermodynamic and Kinetic Selectivity

It is highly desirable for molecular devices such as chemo-/biosensors to carry receptor groups (hosts) that show a high preference for binding with a target guest species among other possible guests. Such a requirement leads to the consideration of selectivity in the development of host–guest systems. The assessment of selectivity may be made based on two aspects, thermodynamic and kinetic. To illustrate their meanings, let's use the binding of a host (H) with two different guests (G1 and G2) as an example (see Figure 3.25). **Thermodynamic selectivity** is the ratio of the binding constants for the two competing processes, K_{G1}/K_{G2}. Given that the binding constants reflect the magnitude of the Gibbs energy changes of the two equilibria (i.e., $\Delta G° = -RT\ln K$), the thermodynamic selectivity is associated with the relative Gibbs energy changes ($\Delta\Delta G°$) for the two equilibria.

Kinetic selectivity has a very different meaning than thermodynamic selectivity. Kinetic selectivity is based on the relative rates of two competing binding reactions. A guest that binds with the host at a faster rate is more kinetically favored and therefore affords a greater kinetic selectivity than the slower one. From the

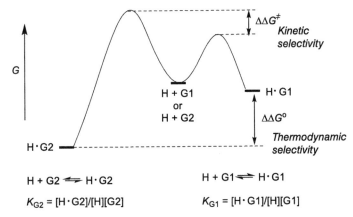

$H + G2 \rightleftharpoons H \cdot G2$

$K_{G2} = [H \cdot G2]/[H][G2]$

$H + G1 \rightleftharpoons H \cdot G1$

$K_{G1} = [H \cdot G1]/[H][G1]$

Figure 3.25 Gibbs energy profiles for two competing binding processes.

thermodynamic viewpoint, the kinetic selectivity reflects the difference between the energy barriers for the two competing binding processes. The transition state of a binding process thus plays a determining role in its kinetic selectivity. A host with a strained geometry that is well preorganized for binding would require a low activation energy and hence would exhibit excellent kinetic selectivity. In certain cases, the host needs to undergo desolvation prior to binding with a guest, so the solvent effect also plays a significant role in the kinetics of host–guest binding.

3.6 Common Scaffolds of Synthetic Receptors

3.6.1 Crown Ethers

Crown ethers are a very important class of receptors (hosts) for various metal cations and cationic guests. Since the initial discovery of dibenzo[18]crown-6 by Pederson in 1967, a large number of crown ethers and crown ether variants have been synthesized and investigated. Figure 3.26 lists the general examples of unfunctionalized crown ethers and their selectivity for different alkali metal cations. Crown ethers can be named as [m]crown-n, in which m is the number of atoms on the ring skeleton and n is the number of ligand (oxygen) atoms. For example, [18]crown-6 has a cavity size of 2.6–3.2 Å and is one of the most widely used crown ethers for binding with alkali and alkaline earth metal cations. Typically, a crown ether prefers a nearly planar conformation to host a metal ion that well matches its cavity; for instance, the binding of [18]crown-6 with K^+ ion illustrated in Figure 3.18. When the size of the metal cation deviates from the ideal size, the binding still occurs but the conformation of the crown ether takes on a more twisted shape. As such, the binding strength is decreased. Such trends can be clearly seen from Table 3.1, where the binding constants of crown ethers

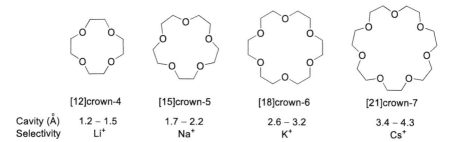

	[12]crown-4	[15]crown-5	[18]crown-6	[21]crown-7
Cavity (Å)	1.2 – 1.5	1.7 – 2.2	2.6 – 3.2	3.4 – 4.3
Selectivity	Li^+	Na^+	K^+	Cs^+

Figure 3.26 Structures of common crown ethers and their size complementarity to different alkali metal cations.

Table 3.1 Binding constants (log K, M^{-1}) for selected crown ethers and alkali metal ions measured in methanol at 20 °C.

Crown ether	Na^+	K^+	Rb^+	Cs^+
[12]crown-4	1.70	1.30	–	–
[15]crown-5	3.24	3.43	–	2.18
[18]crown-6	4.35	6.08	5.32	4.70
[21]crown-7	2.52	2.35	–	5.02

with various cations are listed. For example, the largest binding constant involving [18]crown-6 in methanol is with K^+ as opposed to other smaller or larger alkali metal cations, which well reflects the factor of size complementarity.

Crown ethers with different ring sizes afford cavities that can be used as selective ligands for different cations according to their sizes. However, in reality the size-match criterion should be treated with care, especially in the cases where solvent effects are significant. It is well known that the hydration diameter of a metal cation in an aqueous solvent is very different from the actual size of the ion. For example, the hydrated diameter of lithium ion is much larger than that of sodium ion. So, even though a small crown ether such as [12]crown-4 shows better size complementarity with lithium ion than with sodium ion, it binds more strongly with sodium than lithium in an aqueous solution. Taking advantage of this selectivity, Warnock et al. in 2021 prepared a cross-linked polymer membrane through Ru-catalyzed ROMP reactions [19]. As shown in Figure 3.27, the membrane contains [12]crown-4 as a selective ionophore and hence shows a very high transport selectivity for lithium ion over sodium ion.

The backbone of crown ethers can also be modified with cyclic moieties, such as benzo, cyclohexyl, and binaphthyl units, so that the crown ether structures would attain a sufficiently large cavity for binding with large organic cations or would exhibit a specific stereoconfiguration for chiral recognition and catalysis (Figure 3.28A).

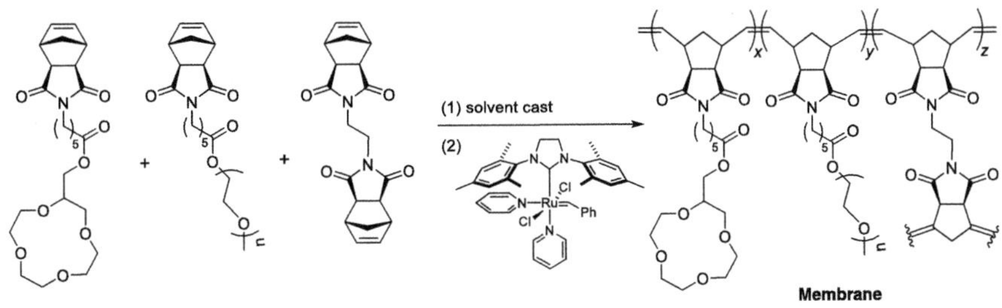

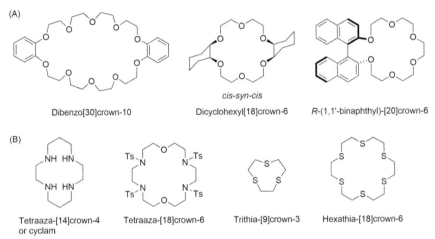

Figure 3.27 Preparation of a [12]crown-4 functionalized polymer membrane that shows a high permeability selectivity for lithium ion.

(A)

Dibenzo[30]crown-10   Dicyclohexyl[18]crown-6   *R*-(1,1'-binaphthyl)-[20]crown-6

cis-syn-cis

(B)

Tetraaza-[14]crown-4   Tetraaza-[18]crown-6   Trithia-[9]crown-3   Hexathia-[18]crown-6
or cyclam

Figure 3.28 Variants of crown ethers: (A) benzo, cyclohexyl, and binaphthyl crown ethers, (B) examples of azacrown and thiacrown ethers.

Replacement of the oxygen atoms in a crown ether with heteroatoms such as nitrogen or sulfur atoms leads to aza- and thia-crown ethers (Figure 3.28B). Compared with typical crown ethers, azacrown and thiacrown ethers show very different affinities, geometries, and binding stoichiometry with various cations. Generally speaking, nitrogen is more basic than oxygen and hence serves as a better ligand for transition metal ions. Alkylation of the nitrogen in an azacrown ether can exert a significant effect on the basicity, which in turn influences the binding properties. Moreover, the nitrogen center is trivalent, allowing the crown ether structure to be further extended to more complex 3D structures (e.g., lariat ethers and cryptands). Sulfur is a much larger heteroatom than oxygen and shows a significant degree of softness. As a result, thiacrown ethers usually show favorable binding with soft cations, such as silver and gold ions. The development of numerous crown ether derivatives has greatly enriched the toolbox for generating cation-selective ligands and **ionophores**, which are compounds capable of forming complexes with specific ions and facilitating their transport across cell membranes.

3.6.2 Podands and Lariat Ethers

The term **podand** was first introduced by Vögtle and Weber in 1979 [20], referring to open-chain ligands of crown ether type. One year later, they came up with a nomenclature method to systematically name neutral ligands based on their topologies (Figure 3.29). In this naming method, acyclic ligands carrying donor atoms are called **podands**. Multidentate macrocyclic ligands, such as crown ethers, belong to the class of **corands**. Ligands taking 3D spherical (cage-like) structures are termed **cryptands**.

Like crown ethers, podand hosts can bind with cations but the binding strength is relatively weak due to the lack of preorganization. On the other hand, the flexibility of podand hosts allows them to engage with cation guests through various binding modes, including helical and bridging modes. If a podand is terminated with a rigid functionality (e.g., an arene, see Figure 3.30), the binding can be enhanced as a result of its rigidified end groups.

Crown ethers or similar macrocyclic ligands that contain one or more podand arms to gain enhanced binding with cation guests are called **lariat ethers** [21]. The name comes from the Spanish *la reata*, which means "the rope." Figure 3.31 compares the binding properties of a group of lariat ethers derived from aza-[18] crown-6 ether. The results clearly show that the side arm attached to the nitrogen atom of the crown ether participates in the binding with metal ions and therefore exerts significant influence on the binding affinity and selectivity.

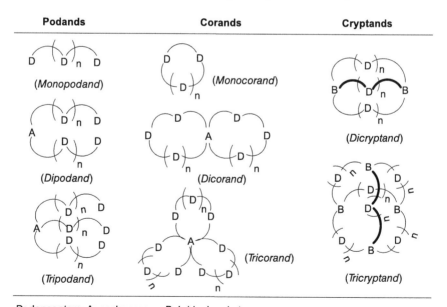

D: donor atom; A: anchor group; B: bridgehead atom

Figure 3.29 Topology and classification of neutral ligands devised by Vögtle and Weber.

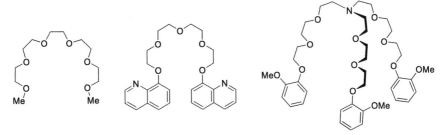

Figure 3.30 Examples of podand ligands without and with rigidified end groups.

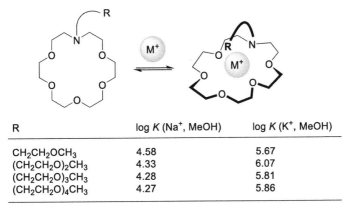

R	log K (Na$^+$, MeOH)	log K (K$^+$, MeOH)
CH$_2$CH$_2$OCH$_3$	4.58	5.67
(CH$_2$CH$_2$O)$_2$CH$_3$	4.33	6.07
(CH$_2$CH$_2$O)$_3$CH$_3$	4.28	5.81
(CH$_2$CH$_2$O)$_4$CH$_3$	4.27	5.86

Figure 3.31 Comparison of the binding properties of lariat ethers with varying side arms.

Functionalization of the two nitrogen sites in a diazacrown ether with appendant groups leads to a type of bibracchial lariat ethers, which were termed BiBLE hosts by Gokel and co-workers [22]. A broad range of BiBLE lariat ethers has been developed to show selective binding for various metal cations. Figure 3.32 lists a group of *N,N'*-disubstituted 4,13-diaza-[18]crown-6 derivatives and their binding constants for complexation with sodium cation. The attachment of benzyl groups to the crown ether was intended to introduce cation-π interactions to reinforce the cation binding of the BiBLE ligands.

3.6.3 Spherands, Hemispherands, and Cryptaspherands

As discussed in Section 3.4, spherands are a class of rigid and highly preorganized macrocyclic hosts which contain *meta*-bridged phenol groups (Figure 3.33). Spherands were initially designed by Cram and co-workers, showing very strong and selective binding affinity for lithium and sodium cations. Relatively large cations, however, cannot fit into the rigid binding pocket of a spherand; therefore,

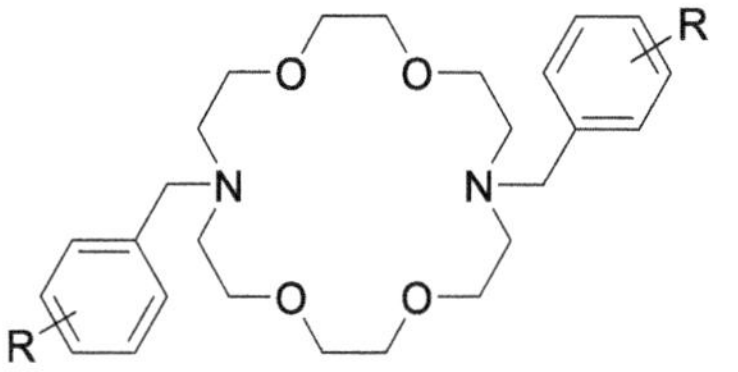

R	log K (Na$^+$, MeOH)
H	2.68
4-OMe	2.79
4-Cl	2.40
4-CN	2.07
2-OH	2.03

Figure 3.32 Dibenzyl-substituted BiBLE ligands and their binding with sodium cation.

Figure 3.33 Examples of spherands designed by Cram.

their binding with a spherand is negligible. Furthermore, the oxygen atoms in a spherand can also be tethered to alter the conformation of the phenyl rings.

The rigid structures of spherands make their binding processes considerably slower than those of crown ethers. Decomplexation of the metal cation from a spherand-metal cation complex is also very slow. Modifications of the spherand structure with flexible moieties, such as podands and crown ethers, have led to hybrid ligands called **hemispherands** and **cryptaspherands**. Figure 3.34 illustrates the examples of a hemishperand and a cryptaspherand. Incorporation of podand and crown ether groups into such structures results in hybrid ligands that show binding for various metal cations. The selectivity of these ligands for different metal cations is controlled through tailoring their molecular backbones and tether groups.

3.6.4 Calixarenes and Resorcinarenes

Calixarenes are a class of macrocyclic compounds with different numbers of phenolic units linked by methylene bridges at the *ortho* positions [23]. These compounds are named because of the resemblance of their molecular shapes to that of a chalice or a vase (*calix crater* in Greek). As discussed in the previous chapter (see Section 2.3.6), calixarenes can be synthesized through a condensation reaction between a 4-substituted phenol and formaldehyde. By varying the reaction

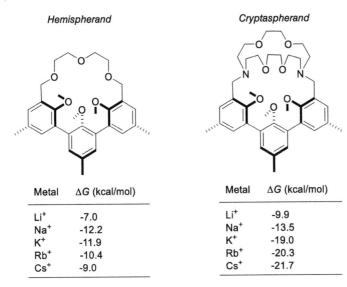

Metal	ΔG (kcal/mol)
Li$^+$	-7.0
Na$^+$	-12.2
K$^+$	-11.9
Rb$^+$	-10.4
Cs$^+$	-9.0

Metal	ΔG (kcal/mol)
Li$^+$	-9.9
Na$^+$	-13.5
K$^+$	-19.0
Rb$^+$	-20.3
Cs$^+$	-21.7

Figure 3.34 Structures of a hemispherand and a cryptaspherand and their binding Gibbs energies for alkali metal cations under standard conditions.

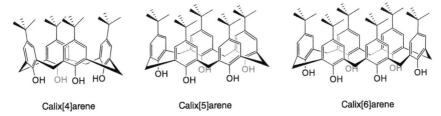

Figure 3.35 Calix[*n*]arenes with *t*-butyl groups substituted on the upper rim.

conditions (e.g., temperature, base, and ratio of reactants), a range of calix[*n*] arenes can be produced, in which the number of repeat unit *n* varies from 4 to 16. However, the commonly encountered calix[*n*]arenes are those with *n* = 4, 6, and 8. Figure 3.35 illustrates selected examples of calixarenes, in which bulky *t*-butyl groups are substituted on the so-called upper rim. The presence of a *t*-butyl group on the phenol starting material is beneficial, since it blocks the reactive *para* position of the phenol, preventing the formation of Bakelite-type polymer products during the calixarene synthesis. The upper rim has wider opening than the opposite lower rim. Usually, the lower rim of a calixarene is functionalized with hydrophilic hydroxy groups that can be further converted into other functional groups to change the structural properties and chemical nature of the calixarene framework, leading to versatile binding properties for cations, anions, and neutral molecules as guests [24].

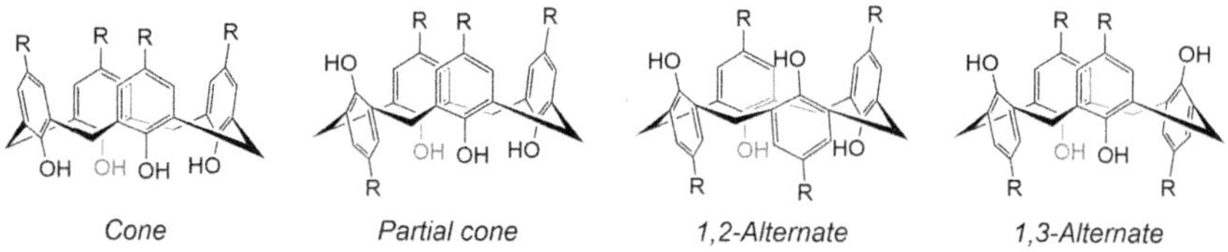

Figure 3.36 Different conformations of a calix[4]arene.

The macrocyclic 3D cavity of calixarenes makes them popular as hosts and receptors in many applications. Calixarenes can adopt a variety of conformations owing to the flexibility of the methylene bridge. For example, calix[4]arene can exist as different conformers in the solution phase, including the cone, partial cone, 1,2-alternate, and 1,3-alternate conformers (see Figure 3.36). All the phenolic hydroxy groups on the lower rim of the rigid cone conformation are closely positioned so that they can form strong hydrogen bonds that enhance the stability of the structure.

Compared with other macrocyclic hosts, calixarenes possess the advantage of a preorganized cavity and are easily tunable at both rims. For these reasons, a vast number of calixarene derivatives have been reported in the literature, showing intriguing applications in molecular recognition, chemo-/biosensing, drug delivery, and so on.

The condensation reactions of resorcinol with aldehydes produce another class of macrocyclic compounds called **calix[*n*]resorcinarenes** or **resorc[*n*]arenes**. Such condensations were performed by Adolf von Baeyer as early as in 1872, who initially obtained an unidentified crystalline product. Following Baeyer's synthesis, numerous efforts were made to elucidate the molecular structure of the product, and it wasn't until 1968 that Erdtman et al. unequivocally proved that the product is a tetrameric metacyclophane by X-ray structural analysis (Figure 3.37) [25]. Studies of the mechanism for the macrocyclization reaction of resorcinol showed that during the condensation the acyclic tetramer undergoes a rapid cyclization, owing to its preorganized structure enabled by hydrogen bonding. As a result, the calix[4]resorcinarene system is highly favored over other possible

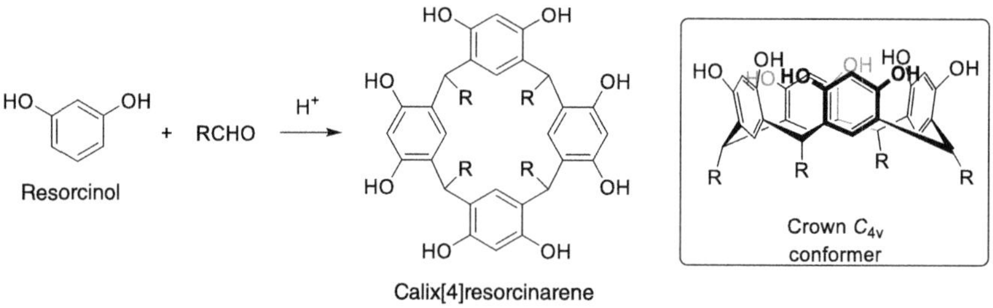

Figure 3.37 Formation of calix[4]resorcinarenes through condensation reactions of resorcinol and aldehydes.

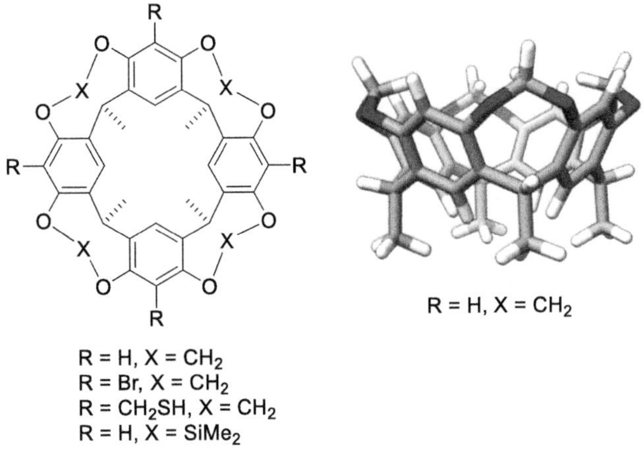

R = H, X = CH$_2$

R = H, X = CH$_2$
R = Br, X = CH$_2$
R = CH$_2$SH, X = CH$_2$
R = H, X = SiMe$_2$

Figure 3.38 Molecular structure of a series of bridged calix[4]resorcinarenes designed by Cram as cavitands (left) and the molecular model of the unsubstituted structure (right).

condensation products as highly dilute conditions or template effects are not required for cyclization.

Like calix[4]arenes, a calix[4]resorcinarene can adopt a multitude of conformations in the solution phase, among which only the crown C_{4v} conformer (see inset of Figure 3.37) exhibits a vase-like molecular shape; however, the cavity is somewhat shallow.

3.6.5 Cavitands and Carcerands

A **cavitand** is a molecular container with an enforced concave surface, which is generally open at one end. A guest molecule can be trapped inside a cavitand to form a host–guest complex, often referred to as a **caviplex** or **cavitate**. Molecular building blocks with an intrinsic curvature can be used to construct cavitand hosts. The first group of cavitands was developed by Cram and co-workers, who utilized the scaffold of calix[4]resorcinarene to create a class of bridged derivatives with a more preorganized and rigidified conformation (Figure 3.38). The rigidified conformation of these bridged resorcinarenes allows them to act as cavitands for specific molecular recognition. The preparation of resorcinarene-based cavitands can be achieved through modifications of the bridge groups that connect the phenolic units, which in turn regulate the shape, dimensions, and complexation properties of the resulting cavity.

With the above cavitands as starting materials, Cram and co-workers carried out the synthesis of a class of **carcerand** hosts by covalently joining two cavitands together through four linkers. Carcerands are closed molecular containers or

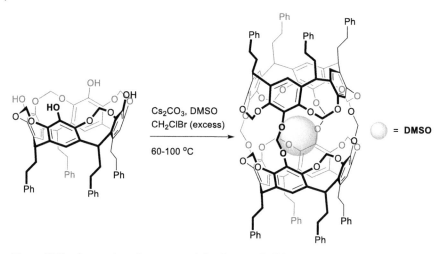

Figure 3.39 Generation of a carcerand that hosts a DMSO molecule in its cavity.

capsules, which are devoid of portals of significant size for guest molecules to enter or leave. A guest molecule when trapped inside a carcerand cannot leave its cavity unless covalent bond cleavage occurs on the carcerand host. Figure 3.39 illustrates an example of the preparation of a methylene-bridged carcerand by Cram [26]. When the shell-closure reaction was performed in DMSO, the resulting carcerand permanently trapped a DMSO inside, forming a **carcerplex**.

The major limitation of carcerands in host–guest chemistry is that the guest cannot escape from the carcerplex unless covalent bond cleavage occurs. To address this issue, **hemicarcerands** have been designed and investigated. There are two approaches to construct a hemicarcerand; one is to use longer and more flexible bridges, and the other is to reduce the number of bridge units (Figure 3.40). Both ways can create relatively larger portals to allow the reversible passage of the guest.

Studies of the binding and kinetic properties involved in the complexation and decomplexation of hemicarcerands and guests led to the discovery of **constrictive binding** [27]. As illustrated in Figure 3.41, intrinsic binding energy is the Gibbs energy difference between a hemicarcerplex and its corresponding free hemicarcerand and guest. The intrinsic binding energy reflects the magnitude of the non-covalent interactions between the guest and the host's inner surface. Constrictive binding, on the other hand, is a kinetic measure that describes the activation energy barrier for a guest to enter or escape the inner cavity of a hemicarcerand through a size-restricting portal on the surface of the host.

Many hemicarcerplexes are very stable because of their large constrictive binding energies, which are useful in separation and isolation techniques. However, their processes of dissociation and/or exchange of guests are kinetically

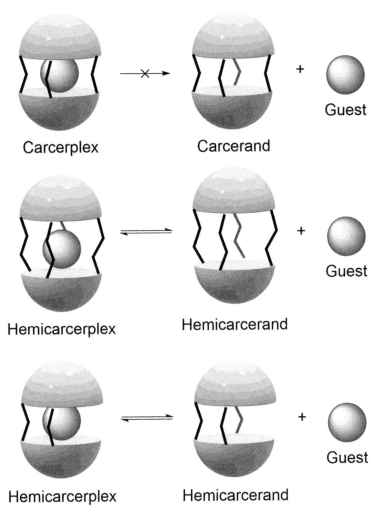

Figure 3.40 Comparison of the structures of a carcerand and two hemicarcerands.

slow and require high temperature. For some applications, a rapid release of the guest is desirable. To meet this need, various types of "gated" hemicarcerplexes have been developed. In such systems, an external stimulus or signal is utilized to trigger the transformation of a carcerand with a small portal to a hemicarcerand with a large enough portal for a guest to pass through. Usually, the closed form (carcerand) has a large constrictive binding energy. When converted into the open form (hemicarcerand), the constrictive binding energy is significantly reduced, allowing rapid guest exchange at ambient temperature. The switching behavior of a gated hemicarcerand can be enabled by means of conformational changes

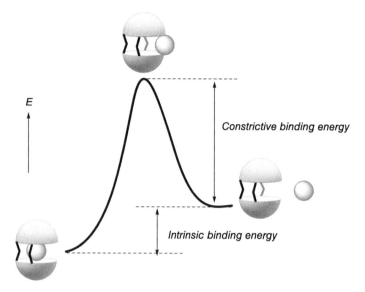

Figure 3.41 Illustration of the constrictive binding energy and intrinsic binding energy.

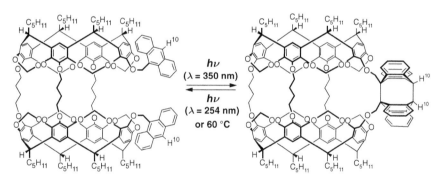

Figure 3.42 Photochemical reversible transformations between the open hemicarcerand and closed carcerand. (Reproduced from *Angew. Chem. Int. Ed.* 2013, *52*, 655–659).

or covalent reactions triggered by photoirradiation, redox reactions, pH changes, complexation with a metal cation, nucleophilic attack, and so forth. Figure 3.42 shows a photochemically gated hemicarcerand designed by Houk and co-workers in 2013 [28]. The system in its open form allows the passage of a guest molecule such as 1,4-dimethoxybenzene to be trapped or released. Under irradiation at 350 nm, the two anthryl groups undergo a photodimerization reaction, forming a rigid covalent linkage to lock the system into a carcerand. Upon irradiation of at a shorter wavelength (254 nm) or heating, the anthracene dimer can undergo dissociation, which subsequently converts the system back to the hemicarcerand.

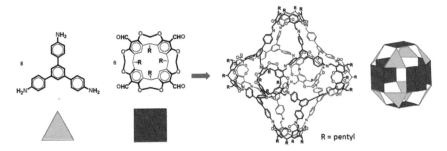

Figure 3.43 Construction of a molecular rhombicuboctahedron through imine condensation reactions. (Reproduced from *Chem. Eur. J.* 2013, *13*, 8953 with modifications).

Through this reversible photochemical reaction, the hemicarcerand is equipped with a "gate" function to encapsulate or release a guest in a controlled manner.

Besides covalently linking two cavitands, a carcerand or hemicarcerand can also be assembled through dynamic covalent chemistry such as olefin metathesis, and exchange reactions involving acetals, oximes, hydrazones, semicarbazones, imines, disulfides, and boronic esters. Moreover, hydrogen-bonding interactions can be utilized for carcerand and hemicarcerand synthesis as well. Since such syntheses are thermodynamically controlled, the resulting systems may go beyond the 1:1 assembly of two cavitands. In some cases, very large nanocapsulates but thermodynamically favored polycavitand assemblies are formed. One impressive example of such chemistry is the nanocapsule synthesized by Warmuth and co-workers in 2007 (shown in Figure 3.43) [29]. Herein a molecular rhombicuboctahedron with a solvodynamic diameter of 3.9 nm is assembled through the imine condensation reactions of six cavitand molecules and eight molecules of a triamine.

3.6.6 Cyclodextrins and Cucurbiturils

Cyclodextrins (CDs) are a family of water-soluble cyclic oligosaccharides obtained from biodegradation of starch with glucanotransferase enzyme [30]. CDs are commonly seen in three major forms, namely α-CD, β-CD, and γ-CD, which contain six, seven, and eight glucose units linked through α-1,4-glycosidic bonds, respectively. These three CDs are also the most widely available CDs. As shown in Figure 3.44, their molecular structures take a conical shape with hydrophilic hydroxy groups located on the outer ring, enabling them to possess good water solubility. The inner cavity of a CD is hydrophobic in nature and its dimensions vary from one CD to another. Larger homologues of CDs are also available, but their structures show more distorted conformations as a result of increasing steric encumbrance caused by large-ring strain.

Figure 3.45 describes the structural parameters for CD homologues with the number of repeating glucose units ranging from 6 to 11. The varying sizes offer tunable nanocarriers for making various inclusion complexes [31]. Owing to their

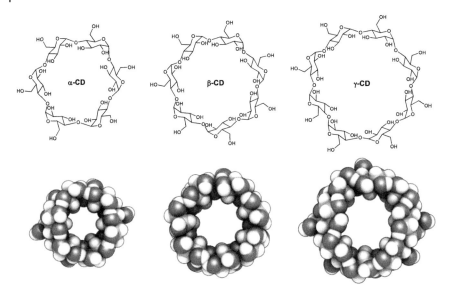

Figure 3.44 Molecular structures and CPK models of α-CD, β-CD, and γ-CD.

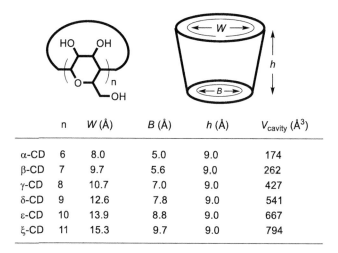

	n	W (Å)	B (Å)	h (Å)	V_{cavity} (Å^3)
α-CD	6	8.0	5.0	9.0	174
β-CD	7	9.7	5.6	9.0	262
γ-CD	8	10.7	7.0	9.0	427
δ-CD	9	12.6	7.8	9.0	541
ε-CD	10	13.9	8.8	9.0	667
ξ-CD	11	15.3	9.7	9.0	794

Figure 3.45 Structural parameters for various CDs.

water solubility, biocompatibility and non-toxicity, CDs have found extensive application in the food industry and in drug formulation.

In supramolecular and nanochemistry, CDs have been frequently utilized as water-compatible hosts for making inclusion complexes and functional nanoassemblies. Figure 3.46 illustrates a CD-based rotaxane designed by Smith and

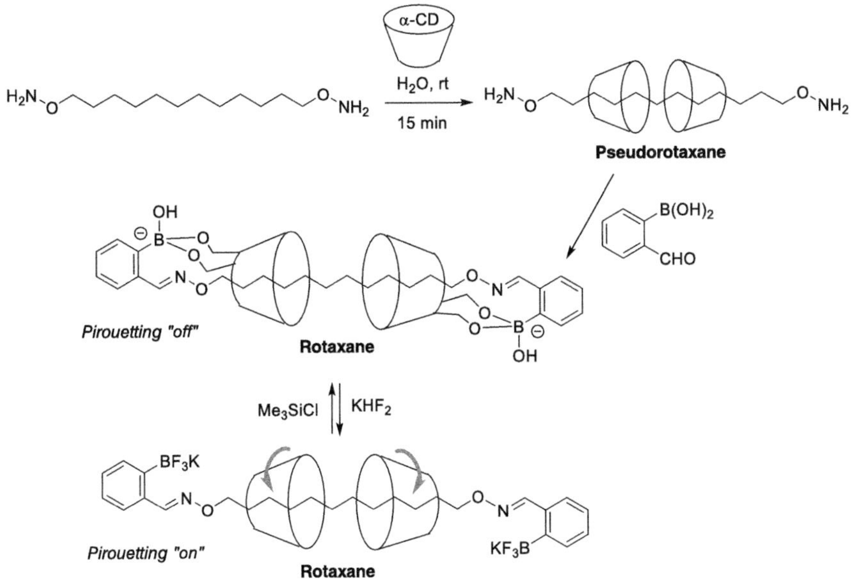

Figure 3.46 Synthesis of an α-CD-based rotaxane system showing the performance of switchable pirouetting.

co-workers in 2018, which shows the ability to perform switchable pirouetting [32]. At first, two α-CD molecules were readily complexed with a 12-carbon alkyl chain to form a pseudorotaxane. The assembly of this pseudorotaxane is driven by the hydrophobic interactions between the alkane and the inner cavity of the α-CD. The terminal positions of the pseudorotaxene were next end capped with two phenylboronic acid stoppers through condensation reactions to form a rotaxane, in which the hydroxy groups on the rims of the two α-CDs are bonded to the boron centers to form boronate ester groups. In this way, the two α-CD wheels are restricted from rotation around the 12-carbon axle (i.e., pirouetting "off" state). Treatment of the rotaxane with KHF_2 led to the cleavage of the boronate ester linkages, allowing the two α-CD wheels to undergo free rotation (pirouetting "on" state). Remarkably, the pirouetting "on" state can be switched into the "off" state through reaction with trimethylsilyl chloride ($SiMe_3Cl$), which converts the trifluoroborate group back to boronic acid and then complexed with the hydroxy groups on the α-CD rims.

Like CDs, **cucurbit[*n*]urils** (CB[*n*]) constitute another popular class of water-soluble macrocyclic hosts, which have been widely used to generate functionalizable molecular containers [33]. The properties of CBs have been briefly discussed in the previous chapter (see Section 1.4.6). CBs are composed of glycoluril units that are assembled from an acid-catalyzed condensation reaction of glycoluril and

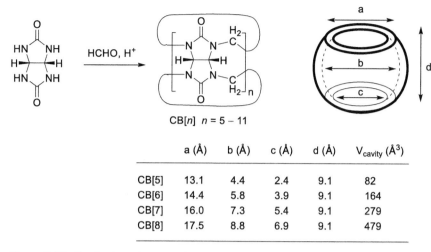

	a (Å)	b (Å)	c (Å)	d (Å)	V_{cavity} (Å^3)
CB[5]	13.1	4.4	2.4	9.1	82
CB[6]	14.4	5.8	3.9	9.1	164
CB[7]	16.0	7.3	5.4	9.1	279
CB[8]	17.5	8.8	6.9	9.1	479

Figure 3.47 Synthesis and structural parameters of selected CB[n] homologues.

formaldehyde (Figure 3.47). Reaction of glycoluril with formaldehyde in a mineral acid (e.g., 9 M H_2SO_4 or conc. HCl) yields a mixture of CB[n] homologues, which can be separated using different methods such as crystallization, dissolution, and chromatography. CB[6] was first reported by Behrend and co-workers in 1905 [34], but its chemical nature and structure were not unraveled until 1981 when Mock et al. conducted a full characterization analysis [35]. CB[n] molecules exhibit a pumpkin-like shape, which is the reason why they are named cucurbiturils. Each CB has a hydrophobic cavity and two identical carbonyl-laced portals. The varying cavity and portal sizes exhibited by the CB[n] family make them useful as supramolecular hosts in the application of molecular recognition.

CB[6] has been found to form stable host–guest complexes with various alkyl ammonium ions. CB[7] can form 1:1 complexes with a range of molecules and cations, such as naphthalene derivatives, protonated admantanamine, methyl viologene dication, ferrocene derivatives, and carborane. CB[8] has a cavity large enough for the formation of 1:2 complexes. Remarkably, CB[10] can trap CB[5] inside its cavity. The varying binding properties and selectivity of CBs have been utilized to produce functional nanomaterials and interfacial assemblies. Figure 3.48 illustrates an elegant example of using the host–guest chemistry between CB[7] and ferrocene to create a supramolecular Velcro. In this work, Kim and co-workers prepared CB[7] and ferrocene-based SAMs on silicon surfaces [36]. The complementary host–guest binding between CB[7] and ferrocene enables the two surfaces show strong adhesion underwater. This method offers a supramolecular "Velcro" strategy that is useful for various adhesion applications in aqueous environments.

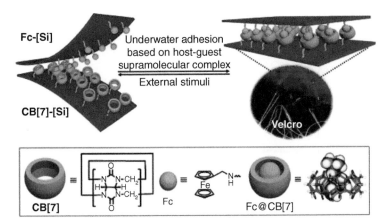

Figure 3.48 Supramolecular Velcro for underwater adhesion between CB[7] and ferrocene-modified surfaces. (Reproduced from *Angew. Chem. Int. Ed.* 2013 / Ahn et.al., 2013 / JOHN WILEY & SONS, INC).

3.7 Templated Synthesis of Macrocycles

So far, we have seen many examples of macrocyclic molecules and oligomers that work as convergent ligands to bind with divergent guests (e.g., metal cations, neutral molecules) through synergistic non-covalent interactions such as metal coordination, hydrogen bonding, electrostatic attraction, and π-stacking. These host–guest interactions not only allow functional supramolecular systems and devices to be developed, but offer useful strategies for organic synthesis. One application of these host–guest interactions is known as the **template synthesis**. Template synthesis is a very broad topic and not so easy to strictly define. In general, it can be referred to any synthetic reaction that takes advantage of a chemical template to organize the assembly of atoms or molecules to achieve particular geometries, shapes, and selectivity. One well-known example of the template synthesis is the replication of DNA, in which each strand of the DNA double helix serves as a template (or mold) to regulate the formation of the second strand. An analogy of template synthesis can be made on the construction of a stone arch as illustrated in Figure 3.49. Tapered stone called **voussoirs** are first placed on an arch-shaped wood framework called **centering** or **falsework**. The arch only holds together once the final voussoir, which is the **keystone** at the crown of the arch, is in position. At the last stage, the centering is removed from the stonework.

For a synthetic chemist, the implementation of a template synthesis is quite similar to the mason work mentioned above. In the first step, a template center, which could be a metal cation, an anion, or a neutral molecule, interacts with a number of ligands in order to orient them in a desired arrangement. The reactive

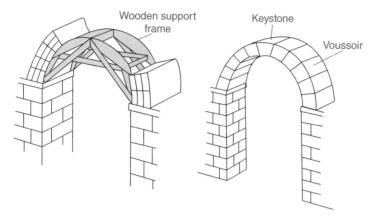

Figure 3.49 Construction of a stone arch on a temporary wood framework.

sites of the ligands are brought to proximity so they can react to form covalent linkages. After the reactions are complete, the template can be dissociated and ready to be reused in the next synthetic cycle. Figure 3.50 shows an example of a template S_N2 reaction. Here the substitution step takes place after the association of an amine and an organic bromide substrate occurs with a template molecule through hydrogen bonding.

Many macrocyclization reactions have been performed using a template synthetic strategy, including the serendipitous discovery of dibenzo[18]crown-6 ether by Pederson, in which sodium cations play an important role. A macrocyclization without using a temple synthetic strategy usually affords the desired cyclic product in a low yield, owing to the competing side reactions that lead to a statistical distribution of cyclic and acyclic oligomeric products (Figure 3.51A). High dilution conditions can be applied to attenuate the formation of acyclic side products, but this strategy is neither ideal for large-scale synthesis nor sufficient to address the size selectivity for the macrocyclization. In a template synthesis as illustrated in Figure 3.51B, a certain number of acyclic precursors are at first bound to a multidentate template to form a defined complex, in which the precursors are predisposed to react by following only one macrocyclization pathway. Once the macrocycle is assembled, the template can be removed under certain conditions. Obviously, the template effect makes the desired macrocyclization far more kinetically favored than other possible reaction pathways.

Figure 3.52 shows some classical examples of the template effects of various cations in the synthesis of crown ethers and related macrocycles. In these reactions, podand precursors are at first complexed with various cations. The resulting host–guest complexes show geometries that are close to the shapes of

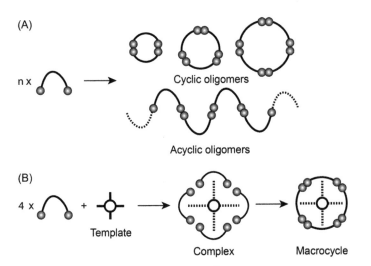

Figure 3.50 An S$_N$2 reaction facilitated by a hydrogen-bonding template.

(A)

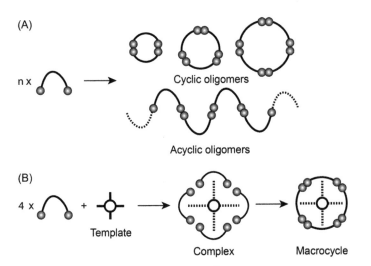

Figure 3.51 Schematic illustrations of macrocyclization (A) without and (B) with a template effect.

desired crown ether products. Subsequent S$_N$2 reactions occur in a regioselective manner to close up the rings. As such, the sizes of crown ether products are controlled by cation templates with different sizes.

Many anions show hydrogen-bonding interactions with suitable ligands. Such binding interactions can also be utilized in the synthesis of macrocycles.

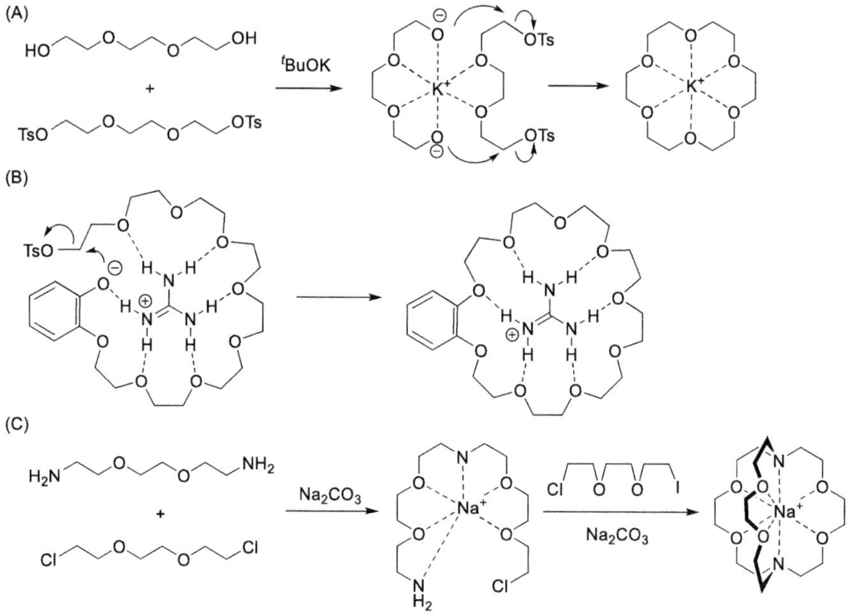

(A)

(B)

(C)

Figure 3.52 Exemplar template synthesis of crown ethers and cryptands.

Figure 3.53 illustrates an elegant example of anion-templated macrocyclization [37]. In this work, Sessler and co-workers investigated the condensation reaction between a diamine ligand and a bispyrrole-dialdehyde in different acidic media. A variety of acids, including HCl, HBr, CH_3COOH, CF_3COOH, H_3PO_4, H_2SO_4, and HNO_3, were employed to promote the condensation reactions, giving different distributions of acyclic oligomers and macrocycles depending on the acid present in the reaction. It is remarkable to note that when H_2SO_4 was used, the condensation reaction yielded a [2+2] macrocycle in nearly quantitative yield. The uniquely high selectivity for this macrocycle can be explained by the template effect depicted in Figure 3.53, where two sulfate anions are bound to the four iminium sites of the macrocycle through hydrogen bonding. After the H_2SO_4 promoted condensation, the free macrocycle could be obtained by addition of an organic base, triethylamine, to break apart the hydrogen-bonding assembly. Titration experiments conducted on the free macrocycle indicated that it shows specific selectivity for tetrahedral anions with large binding constants, such as sulfate ($K_a = 63,500 \pm 3000$ M^{-1}) and dihydrogen phosphate ($K_a = 107,500 \pm 9600$ M^{-1}).

In modern synthetic chemistry, template synthesis has been widely adopted in constructing sophisticated nanostructures, such as rotaxanes, catenanes,

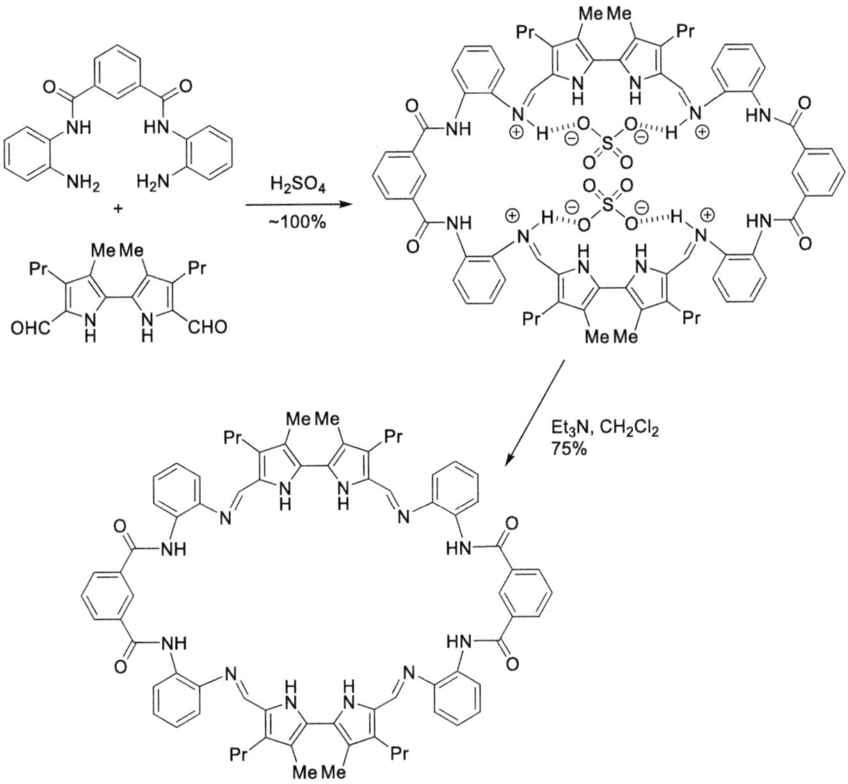

Figure 3.53 A sulfate anion-templated [2+2] macrocyclization.

and large 3D molecular assemblies. An extraordinary example was reported by Anderson's group in 2020, which showcases the power of template synthesis [38]. In their work, a series of giant molecular wheels were successfully synthesized with the aid of predesigned pyridyl ligands as template scaffolds. Figure 3.54 highlights the structure of an amazing macrocyclic compound which contains 12 porphyrin units linked through butadiyne linkages. The assembly of such a giant molecular wheel was enabled by two spoke-like template molecules, which are end capped with pyridyl groups to coordinate with the zinc centers of the porphyrin units. This nanoscale molecular wheel has totally 162 π-electrons delocalized along its backbone. It is also remarkable that ring current analysis of this system gave results in agreement with the prediction by simple Hückel's rule.

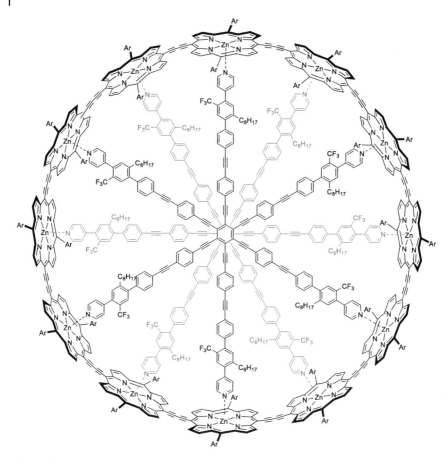

Figure 3.54 A giant molecular wheel prepared through a template synthesis.

Further Reading

- Steed, J. W.; Turner, D. R.; Wallace, K., Core Concepts in Supramolecular Chemistry and Nanochemistry. John Wiley & Sons: 2007.
- Steed, J. W.; Atwood, J. L., Supramolecular Chemistry. 3rd ed.; John Wiley & Sons: 2022.
- Cragg, P. J., A Practical Guide to Supramolecular Chemistry. Jonh Wiley & Son Ltd: West Sussex, 2005.
- Schalley, C. A., Analytical Methods in Supramolecular Chemistry. 2nd ed.; John Wiley & Sons: Weinheim, 2012; Vol. 1.

References

1 Pedersen, C. J., The Discovery of Crown Ethers (Noble Lecture). *Angew. Chem. Int. Ed. Engl.* **1988**, *27*, 1021–1027.

2 Lehn, J., Cryptates: Inclusion Complexes of Macropolycyclic Receptor Molecules. *Pure Appl. Chem.* **1978**, *50*, 871–892.

3 Thordarson, P., Determining Association Constants from Titration Experiments in Supramolecular Chemistry. *Chem. Soc. Rev.* **2011**, *40*, 1305–1323.

4 Benesi, H. A.; Hildebrand, J. H., A Spectrophotometric Investigation of the Interaction of Iodine with Aromatic Hydrocarbons. *J. Am. Chem. Soc.* **1949**, *71*, 2703–2707.

5 Scatchard, G., The Attractions of Proteins for Small Molecules and Ions. *Ann. NY Acad. Sci.* **1949**, *51*, 660–672.

6 Brachvogel, R.-C.; Maid, H.; Von Delius, M., NMR Studies on Li+, Na+ and K+ Complexes of Orthoester Cryptand o-Me2-1.1.1. *Int. J. Mol. Sci.* **2015**, *16*, 20641–20656.

7 Ghai, R.; Falconer, R. J.; Collins, B. M., Applications of Isothermal Titration Calorimetry in Pure and Applied Research—Survey of the Literature from 2010. *J. Mol. Recogn.* **2012**, *25*, 32–52.

8 Le, V. H.; Yanney, M.; McGuire, M.; Sygula, A.; Lewis, E. A., Thermodynamics of Host–Guest Interactions between Fullerenes and a Buckycatcher. *J. Phys. Chem. B* **2014**, *118*, 11956–11964.

9 von Krbek, L. K. S.; Schalley, C. A.; Thordarson, P., Assessing Cooperativity in Supramolecular Systems. *Chem. Soc. Rev.* **2017**, *46*, 2622–2637.

10 Martell, A. E., The Chelate Effect. In *Werner Centennial*, Kauffman, G. B., Ed. American Chemical Society: **1967**; Vol. *62*, pp 272–294.

11 (A) Badjić, J. D.; Nelson, A.; Cantrill, S. J.; Turnbull, W. B.; Stoddart, J. F., Multivalency and Cooperativity in Supramolecular Chemistry. *Acc. Chem. Res.* **2005**, *38*, 723–732; (B) Fasting, C.; Schalley, C. A.; Weber, M.; Seitz, O.; Hecht, S.; Koksch, B.; Dernedde, J.; Graf, C.; Knapp, E.-W.; Haag, R., Multivalency as a Chemical Organization and Action Principle. *Angew. Chem. Int. Ed.* **2012**,, 10472–10498.

12 Cacciapaglia, R.; Di Stefano, S.; Mandolini, L., Effective Molarities in Supramolecular Catalysis of Two-Substrate Reactions. *Acc. Chem. Res.* **2004**, *37*, 113–122.

13 Cabbiness, D. K.; Margerum, D. W., Macrocyclic Effect on the Stability of Copper(II) Tetramine Complexes. *J. Am. Chem. Soc.* **1969**, *91*, 6540–6541.

14 Wittenberg, J. B.; Isaacs, L., Complementarity and Preorganization. In *Supramolecular Chemistry: From Molecules to Nanomaterials*, Steed, J. W. and Gale, P. A., Ed. John Wiley & Sons, Ltd.: **2012**.

15 Trueblood, K. N.; Knobler, C. B.; Maverick, E.; Helgeson, R. C.; Brown, S. B.; Cram, D. J., Spherands, the First Ligand Systems Fully Organized during

Synthesis Rather than during Complexation. *J. Am. Chem. Soc.* **1981**, *103*, 5594–5596.

16 Cram, D. J.; Lein, G. M., Host-Guest complexation. 36. Spherand and Lithium and Sodium Ion Complexation Rates and Equilibria. *J. Am. Chem. Soc.* **1985**, *107*, 3657–3668.

17 Kyba, E. P.; Gokel, G. W.; De Jong, F.; Koga, K.; Sousa, L. R.; Siegel, M. G.; Kaplan, L.; Sogah, G. D. Y.; Cram, D. J., Host-Guest Complexation. 7. The Binaphthyl Structural Unit in Host Compounds. *J. Org. Chem.* **1977**, *42*, 4173–4184.

18 Rowan, S. J.; Hamilton, D. G.; Brady, P. A.; Sanders, J. K. M., Automated Recognition, Sorting, and Covalent Self-Assembly by Predisposed Building Blocks in a Mixture. *J. Am. Chem. Soc.* **1997**, *119*, 2578–2579.

19 Warnock, S. J.; Sujanani, R.; Zofchak, E. S.; Zhao, S.; Dilenschneider, T. J.; Hanson, K. G.; Mukherjee, S.; Ganesan, V.; Freeman, B. D.; Abu-Omar, M. M.; Bates, C. M., Engineering Li/Na Selectivity in 12-Crown-4–Functionalized Polymer Membranes. *Pnas* **2021**, *118*, e2022197118.

20 Vögtle, F.; Weber, E., Multidentate Acyclic Neutral Ligands and Their Complexation. *Angew. Chem. Int. Ed. Engl.* **1979**, *18*, 753–776.

21 Gokel, G. W.; Barbour, L. J.; Ferdani, R.; Hu, J., Lariat Ether Receptor Systems Show Experimental Evidence for Alkali Metal Cation–π Interactions. *Acc. Chem. Res.* **2002**, *35*, 878–886.

22 Gokel, G. W.; Arnold, K. A.; Delgado, M.; Echeverria, L.; Gatto, V. J.; Gustowski, D. A.; Hernández, J.; Kaifer, A. E.; Miller, S. R.; Echegoyen, L. A., Lariat Ethers: From Cation Complexation to Supramolecular Assemblies. *Pure Appl. Chem.* **1988**, *60*, 461–465.

23 Gutsche, C. D., The Calixarenes. In *Host Guest Complex Chemistry/Macrocycles: Synthesis, Structures, Applications*, Vögtle, F.; Weber, E., Eds. Springer: Berlin, **1985**; pp 375–421.

24 Vicens, J.; Harrowfield, J. Eds., *Calixarenes in the Nanoworld*. Dordrecht, Springer: **2007**.

25 Erdtman, H.; Högberg, S.; Abrahamsson, S.; Nilsson, B., Cyclooligomeric Phenol-Aldehyde Condensation Products I. *Tetrahedron Lett.* **1968**, *9*, 1679–1682; (b) Nilsson, B., The Crystal and Molecular Structure of the Synthetic Tetramer C84H84Br4O16. *Acta Chem. Scand.* **1968**, 732–747.

26 Sherman, J. C.; Cram, D. J., Carcerand Interiors Provide a New Phase of Matter. *J. Am. Chem. Soc.* **1989**, *111*, 4527–4528.

27 Quan, M. L. C.; Cram, D. J., Constrictive Binding of Large Guests by a Hemicarcerand Containing Four Portals. *J. Am. Chem. Soc.* **1991**, *113*, 2754–2755.

28 Wang, H.; Liu, F.; Helgeson, R. C.; Houk, K. N., Reversible Photochemically Gated Transformation of a Hemicarcerand to a Carcerand. *Angew. Chem. Int. Ed.* **2013**, *52*, 655–659.

29 Liu, Y.; Liu, X.; Warmuth, R., Multicomponent Dynamic Covalent Assembly of a Rhombicuboctahedral Nanocapsule. *Chem. Eur. J.* **2007**, *13*, 8953–8959.

30 Crini, G., Review: A History of Cyclodextrins. *Chem. Rev.* **2014**, *114*, 10940–10975.

31 Lai, W.-F.; Rogach, A. L.; Wong, W.-T., Chemistry and Engineering of Cyclodextrins for Molecular Imaging. *Chem. Soc. Rev.* **2017**, *46*, 6379–6419.

32 Zhang, Q.-W.; Zajíček, J.; Smith, B. D., Cyclodextrin Rotaxane with Switchable Pirouetting. *Org. Lett.* **2018**, *20*, 2096–2099.

33 Kim, K.; Selvapalam, N.; Ko, Y. H.; Park, K. M.; Kim, D.; Kim, J., Functionalized Cucurbiturils and Their Applications. *Chem. Soc. Rev.* **2007**, *36*, 267–279.

34 Behrend, R.; Meyer, E.; Rusche, F. I., Ueber Condensationsproducte aus Glycoluril und Formaldehyd. *Justus Liebigs Ann. Chem.* **1905**, *339*, 1–37.

35 Freeman, W. A.; Mock, W. L.; Shih, N. Y., Cucurbituril. *J. Am. Chem. Soc.* **1981**, *103*, 7367–7368.

36 Ahn, Y.; Jang, Y.; Selvapalam, N.; Yun, G.; Kim, K., Supramolecular Velcro for Reversible Underwater Adhesion. *Angew. Chem. Int. Ed.* **2013**, *52*, 3140–3144.

37 Katayev, E. A.; Pantos, G. D.; Reshetova, M. D.; Khrustalev, V. N.; Lynch, V. M.; Ustynyuk, Y. A.; Sessler, J. L., Anion-Induced Synthesis and Combinatorial Selection of Polypyrrolic Macrocycles. *Angew. Chem. Int. Ed.* **2005**, *44*, 7386–7390.

38 Rickhaus, M.; Jirasek, M.; Tejerina, L.; Gotfredsen, H.; Peeks, M. D.; Haver, R.; Jiang, H.-W.; Claridge, T. D. W.; Anderson, H. L., Global Aromaticity at the Nanoscale. *Nat. Chem.* **2020**, *12*, 236–241.

4

Chemistry of Carbon Nanoallotropes

4.1 Classification of Carbon Nanoallotropes

Organic compounds are defined as compounds based on carbon. Historically, the term *organic* was first introduced in 1807 by Jöns Jakob Berzelius to classify materials that are derived from living organisms. Scientists in the early 18th century believed that organic materials contain an immeasurable "vital force." Compounds derived from minerals, on the other hand, were called inorganic since they lack such vital force. Nowadays, the definition of organic compounds has gone far beyond the original idea and understanding. The surprising observation made by Friedrich Wöhler in 1828 that urea, a compound excreted by mammals, could be generated from heating ammonium cyanate, an inorganic mineral, marked the beginning of organic chemistry, which deals extensively with the study of carbon-based molecules.

   The ubiquitous presence of carbon in organic compounds can be ascribed to the unique properties of carbon. In the periodic table, carbon is located right in the middle of the second-row elements. As such, carbon can form covalent bonds with elements on either side of its position in the periodic table by sharing its valence electrons with other atoms. As discussed in Chapter 1, in forming a covalent bond, a carbon atom can take on different hybridization states, ranging from sp^3, sp^2 to sp. Carbon–carbon and carbon–hydrogen bonds are the most common bonds found in organic structures. In addition, carbon can form bonds with metals to give diverse organometallic species. Materials solely made up of carbon atoms also widely exist in nature. For example, graphite is the most stable natural form of carbon. It is composed of stacked layers of carbon sheets (known as graphene), where sp^2 carbon atoms are densely packed in a 2D hexagonal lattice. Graphite has been widely used as a lubricant and as a conductor of heat and electricity. Diamond is another common form of carbon found in nature. It is a transparent crystal in which sp^3 hybridized carbons are orderly packed in a diamond cubic crystalline

Organic Nanochemistry: From Fundamental Concepts to Experimental Practice,
First Edition. Yuming Zhao.
© 2024 John Wiley & Sons, Inc. Published 2024 by John Wiley & Sons, Inc.

structure. The properties of diamond are very different from those of graphite. It is an electrical insulator and is the hardest known substance. Furthermore, it exhibits high thermal conductivity and a high refractive index. These properties make diamond a useful material for cutting, polishing, and optical applications. Amorphous carbon is a non-crystalline form of carbon produced from incomplete combustion of material obtained from plants and animal substances. Amorphous carbon includes carbon black and activated carbon, both of which have been extensively used in textiles, plastics, health care, and food industries. Additionally, activated amorphous carbon exhibits a very large specific surface area and high porosity, which are useful properties in adsorption and filtration applications.

In addition to organic compounds and natural carbon materials, carbon has also been found to exist as a range of novel materials, known as **carbon nanoallotropes**. The first of these materials is the C_{60} molecule, which was discovered in 1985 by Harold W. Kroto, Robert Curl, and Richard E. Smalley [1]. C_{60} has a cage-like fused-ring structure that resembles the shape of a soccer ball (Figure 4.1). Nowadays, it is well-known as the **buckyball** or **Buckminster fullerene**, which was named after an American architect, Richard Buckminster Fuller, who is famous for his development of the geodesic dome. The discovery of fullerene not only added a new member to the carbon allotrope family but became a landmark in the development of modern nanoscience and nanotechnology. Kroto, Smalley, and Curl shared the Nobel Prize in Chemistry in 1996 for their discovery of fullerenes.

Several other types of fullerenes, including C_{20}, C_{70}, and even larger fullerenes, were later discovered. It is worth mentioning that C_{20} is the smallest member of the fullerene family, which exhibits a dodecahedral cage structure. It was proposed to be highly unstable and hence is referred to as an unconventional fullerene. Nevertheless, Prinzbach and co-workers in 2000 successfully produced C_{20} fullerene in the gas phase with a lifetime of milliseconds [2]. C_{70} fullerene is a fullerene molecule that consists of 70 carbon atoms and takes a closed, hollow

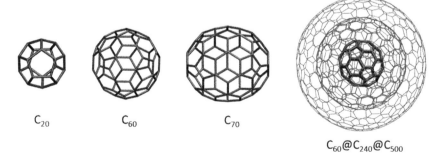

C_{20} C_{60} C_{70}

$C_{60}@C_{240}@C_{500}$

Figure 4.1 Molecular models of C_{20}, C_{60}, and C_{70} fullerenes as well as a multilayer fullerene.

fused-ring structure that looks like a rugby ball. It was reported as the second most prominent carbon cluster, apart from the C_{60} cluster, resulting from the laser vaporization or resistive heating of graphite. In 1990, C_{60} and C_{70} fullerenes were isolated in macroscopic quantities from the soot prepared by the Krätschmer and Huffman protocol [3]. C_{60} and C_{70} fullerenes showed solubility in organic solvents and could be purified by chromatographic separation (e.g., using an alumina column with toluene/hexane mixtures as eluents). Higher fullerenes, such as C_{76}, C_{78}, and C_{84}, were also detected during the analysis of the toluene-soluble soot extract. It is quite remarkable to note that a wide distribution of fullerenes from C_{90} to beyond C_{250} was also found to exist in the crude soot!

Fullerenes are the smallest known carbon nanostructures which exist on the boundary between molecules and nanomaterials. In 1990, Smalley identified that a tubular fullerene must be theoretically achievable, in which two hemispheres of C_{60} fullerene cap the straight segment of a carbon tube with only hexagonal units (sp^2 carbons) in its structure. When Mildred Dresselhaus heard this idea, she named these objects as "buckytubes." Mildred "Millie" Dresselhaus is known as the "Queen of Carbon Science." Her fundamental research on carbon has had a tremendous impact on the development of carbon nanomaterials and nanotechnology. One year later, in 1991 Sumio Iijima, a famous Japanese physicist, discovered **carbon nanotubes** (CNTs). CNTs are tube-shaped carbon nanomaterials made of sp^2 hybridized carbons in fused-hexagon arrangements. Unlike fullerenes, CNTs show diameters of 0.4–40 nm with very large aspect ratios (i.e., length-to-diameter ratios) that can be as high as 132,000,000:1! CNTs can be visualized as seamlessly rolled-up graphene sheets. Depending on the number of graphene layers, CNTs can be described as **single-walled carbon nanotubes** (SWCNTs), **double-walled carbon nanotubes** (DWCNTs), and **multi-walled carbon nanotubes** (MWCNTs) (see Figure 4.2). The structural properties of CNTs make

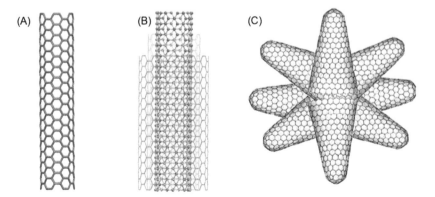

Figure 4.2 Drawings of (A) a single-walled carbon nanotube (SWCNT), (B) a multi-walled carbon nanotube (MWCNT), and (C) single-walled carbon nanohorns (SWCNHs).

them completely different from fullerenes; for example, the tensile strength of CNTs is about 100 times greater than that of steel of the same diameter. CNTs can show metallic or semiconducting behavior depending on their structures. As a synthetic carbon allotrope, CNTs have found extensive applications as composite materials and in the fields of catalysis and nanoelectronics.

Following the discoveries of fullerenes and CNTs, other carbon nanostructures have been developed such as **single-walled carbon nanohorns** (SWCNHs) [4] and **carbon nano onions** (CNOs) [5]. SWCNHs are conical carbon nanostructures constructed from a graphene sheet (Figure 4.2C), while CNOs are made of multiple concentric shells of fullerenes (e.g., the $C_{60}@C_{240}@C_{500}$ assembly in Figure 4.1). The discoveries of these intriguing carbon nanomaterials demonstrate the versatility of carbon to form diverse nanostructures, and their unique shapes and properties have suggested potential applications in many fields [6].

(A)

(B)   (C)

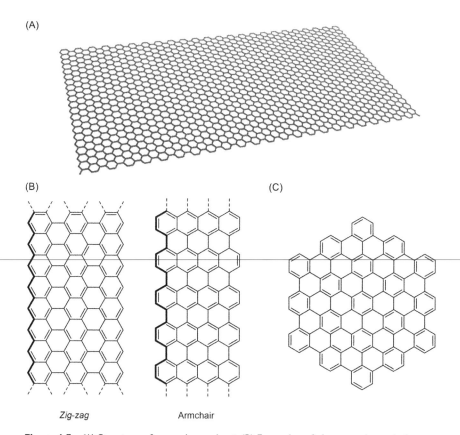

Zig-zag   Armchair

Figure 4.3   (A) Structure of a graphene sheet. (B) Examples of zigzag and armchair graphene nanoribbons. (C) An example of molecular nanographenes.

Another important member in the family of carbon nanomaterials is graphene. Graphene is a carbon allotrope that consists of a single layer of carbon atoms arranged in a 2D honeycomb lattice nanostructure (see Figure 4.3). The existence of graphene was predicted and theoretically explored by Wallace as early as in 1947 [7]. Boehme and co-workers experimentally identified graphene in 1962, and the term **graphene** was introduced by Boehme in 1986 to describe the single layer of graphite [8]. The term graphene is derived from "graphite" and "ene." Graphite designates a mineral and the crystal structure of a modification of the element carbon, while the suffix -ene is used to reflect the sp^2 carbons similar to those of fused aromatic hydrocarbons. In 2004, Andre Geim and Konstantin Novoselov successfully isolated graphene by pulling graphene layers from graphite and then transferring them onto the surface of a silicon wafer, a process called either "micromechanical cleavage" or the "Scotch tape" technique [9]. This technique led to the direct observation of the anomalous quantum Hall effect in graphene, which soon sparked enormous interest in exploring graphene as a new generation of electronic material. For their pioneering work on graphene, Geim and Novoselov received 2010 Nobel Prize in Physics.

In today's material research, graphene and graphene-like nanocarbon materials have been broadly used. To avoid confusion, graphene refers to a 2D carbon sheet with both lateral dimensions exceeding 100 nm (see Figure 4.3A). If one of the dimensions of a carbon sheet is within 1–100 nm, it should be called a **nanographene** [10]. The family of nanographenes includes graphene quantum dots (GQDs), graphene nanoribbons (GNRs), and molecular nanographenes (see Figure 4.3B and C). In recent years, interest in the synthesis and characterization of these novel carbon nanomaterials has grown rapidly, owing to their intriguing structural and optoelectronic properties.

Most of the carbon nanoallotropes briefly mentioned above possess a similar structure of sp^2 carbon atoms arranged in a hexagonal network, accounting for their relatively similar chemical reactivities. On the other hand, their shape, size, and structural differences result in very different physical, electronic, and optical properties. In the following sections, the detailed structural and chemical properties of each class of these carbon nanomaterials are discussed.

4.2 Structural Properties of C_{60} and C_{70} Fullerenes

Although the first fullerene (C_{60}) was experimentally observed in 1985 by Kroto *et al.*, the concept of a C_{60} fullerene molecule was hypothesized fifteen years prior to its discovery. In 1970, Eiji Osawa who was then an assistant professor at Hokkaido University wrote an article entitled "Superaromaticity," in which the icosahedral (I_h) symmetry C_{60} molecule that resembles the shape of a soccer ball was first described [11]. As illustrated in Figure 4.4, the structure of I_h-symmetry C_{60} is

(A)   (B)   (C)

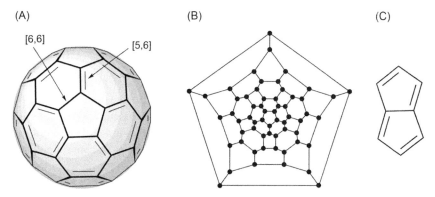

Figure 4.4 (A) Molecular structure of I_h C_{60} fullerene with the [6,6] and [5,6] bonds highlighted, (B) the Schlegel diagram of C_{60} fullerene, and (C) the structure of pentalene.

made up of 60 sp^2 hybridized carbon atoms, exhibiting a hollow polygon shape with 60 vertices and 32 faces. The overall structure of C_{60} is a truncated icosahedron, a geometry long known to mathematicians.

In 1990, an arc discharge method was invented by Krätschmer and co-workers [3], allowing fullerenes to be produced in bulk quantities. Using this method, C_{60} fullerene was isolated in pure form for the first time. In 2003, Frontier Carbon Corporation (Kyushu Japan) set up a facility to generate fullerenes on the scale of 40 tons/year through the combustion method [12] . C_{60} and C_{70} fullerenes are the two most abundant carbon nanoallotropes found in the fullerenes generated by the combustion method. They constitute 60% and 25%, respectively, of the soluble extract from the carbon soot produced. The remaining 15% are higher fullerenes up to C_{96}. So far, C_{60} and C_{70} are the most extensively investigated fullerenes. C_{60} fullerene is a nanoscale spherical-shaped molecule (diameter = 0.7 nm), which represents the highest degree of symmetry of all known molecules. All the carbon atoms in C_{60} are chemically equivalent, giving a single signal at 144 ppm in the ^{13}C NMR spectrum. The topological properties of C_{60} can be depicted by a 2D projection of the 3D structure, called the Schlegel diagram (Figure 4.4B), from which one can clearly see that each pentagon is surrounded by hexagons and there are no two pentagons sharing an edge. This feature is known to comply with the **isolated pentagon rule** (IPR), which describes the most stable isomer(s) of a fullerene. In the structure of a non-IPR fullerene, two pentagons abut to form a pentalene segment. Pentalene (Figure 4.4C) is a highly unstable structure because of its Hückel antiaromaticity (8 π-electrons). It also becomes highly strained when taking a curved shape. Hence, it is highly unlikely to exist in any stable fullerene structure. Actually, C_{60} fullerene is the smallest fullerene that satisfies the IPR.

There are two types of C–C bonds in the C_{60} molecule, namely [6,6] and [5,6] bonds (Figure 4.4A). The [6,6] double bonds connect two adjacent hexagons, showing a bond distance of 1.54 Å, while the [5,6] bonds join a hexagon and a pentagon

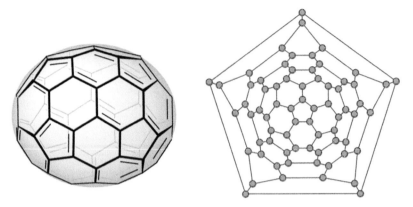

Figure 4.5 Molecular structure of C_{70} fullerene (left) and its Schlegel diagram (right).

with a bond distance of 1.38 Å. The spherical molecular shape of C_{60} makes each carbon atom deviate from the planar configuration that is ideal for sp^2 hybridization. This introduces a large amount of strain energy (10.16 kcal mol^{-1} per carbon atom) to C_{60} and makes it show a tendency to avoid double-bond character within the pentagonal rings. As such, the electron delocalization of C_{60} is limited. In fact, C_{60} fullerene behaves more like a polyene rather than a "superaromatic" molecule. C_{60} fullerene is electron-deficient and can readily react with various electron-rich species. Electrochemically, C_{60} is an excellent electron acceptor. It has been observed that C_{60} underwent six reversible one-electron reductions to form a hexaanion (C_{60}^{6-}). The remarkable electron-accepting properties make C_{60} and its derivatives popular molecular building blocks in the fabrication of advanced organic optoelectronic devices such as organic photovoltaic (OPV) cells.

C_{70} is the second most abundant fullerene as well as the second smallest IPR fullerene. Compared with C_{60}, the structure of C_{70} can be viewed as having ten carbons in the form of an equatorial belt inserted into C_{60}. As a result, C_{70} shows an ellipsoid molecular shape with a D_{5h} symmetry (Figure 4.5). The reduced symmetry makes C_{70} show five chemically equivalent signals in its ^{13}C NMR spectrum. Like C_{60}, C_{70} fullerene is also a useful organic semiconductor with rich photophysical properties. C_{70} can be used as a photo-absorptive material for making high performance organic photovoltaic (OPV) cells or nanocomposites as light responsive catalysts.

4.3 Reactivities of C_{60} Fullerene

4.3.1 Nucleophilic Addition

As discussed above, C_{60} fullerene behaves like an electron-deficient polyene and hence can readily react with various nucleophiles. As illustrated in Figure 4.6, an ionic nucleophile (Nu^-) attacks a C=C bond on the cage of C_{60}, resulting in a

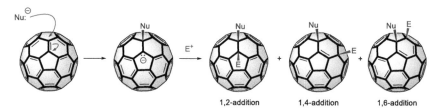

Figure 4.6 General mechanism for the nucleophilic addition reactions on the cage of C_{60} fullerene.

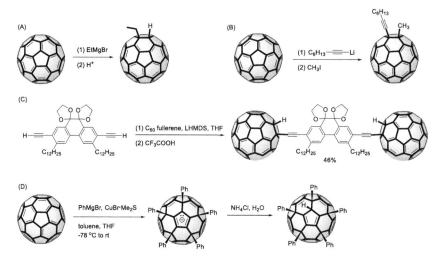

Figure 4.7 Nucleophilic addition reactions of organometallic compounds on the cage of C_{60} fullerene.

fulleride anion intermediate. The intermediate can be further quenched with an electrophile (E^+) to yield a stable neutral product. Owing to the π-delocalized nature of the reactive intermediate, the addition products may show numerous substitution patterns. The generally preferred mode is 1,2-addition, while 1,4-, and 1,6-addition and more extended structures are also possible.

Figure 4.7 illustrates a range of exemplary nucleophilic addition reactions on C_{60} fullerene, in which organometallic nucleophiles directly react with the fullerene cage to form fullerene adducts. As shown in Figures 4.7A and 4.7B, a Grignard reagent or an organolithium compound reacts with excess C_{60} fullerene to form a fulleride anion (RC_{60}^-) intermediate, which can then be stabilized by quenching it with an electrophile such as a protic acid or an alkyl halide. Typically, these reactions result in a 1,2-addition product. If the quenching electrophile is sterically bulky, a 1,4- or 1,6-addition pattern could be generated. It is worth knowing that the nucleophilic addition of a terminal alkyne to C_{60} can be carried out via an *in situ*, one-pot approach as exemplified in Figure 4.7C. Herein, a dialkyne starting material is first treated with lithium hexamethyldisilazide (LHMDS) in the presence of C_{60} fullerene. The resulting alkynyl lithium intermediates can swiftly react with C_{60} to form fulleride adducts. At the end of the reaction, an electrophile (e.g., trifluoroacetic acid or methyl iodide) is added to yield stable products. This method is particularly useful for making mono- and poly-fullerene-substituted π-conjugated systems.

Unlike Grignard and organolithium species, an organocopper compound would attack the cage of C_{60} fullerene at five different sites to afford a stable fullerene cyclopentadienide $R_5C_{60}^-$ intermediate (Figure 4.7D). Nakamura and co-workers first disclosed this type of nucleophilic addition reaction in 1996 by using an organocuprate nucleophile *in situ* generated from a Grignard reagent and CuBr·SMe$_2$ [13]. It is remarkable that this penta-addition reaction on C_{60} is highly regioselective, barely producing any other regioisomers. The detailed mechanism for this penta-addition reaction is described in Figure 4.8, in which Cu(I) plays an important role in driving the reaction toward the final penta-adduct.

C_{60} fullerene can react with primary and secondary aliphatic amines to yield hydroamination products. The products of these reaction are likely to be a mixture

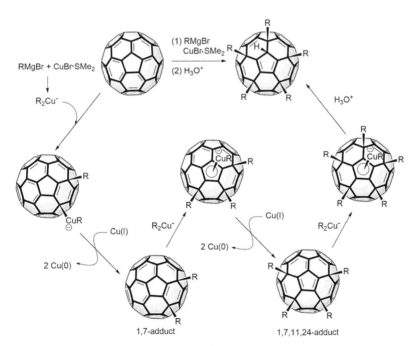

Figure 4.8 Mechanism for the penta-addition of an organocopper reagent on the cage of C_{60} fullerene in a regioselective way.

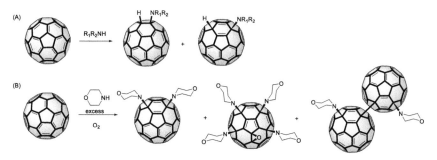

Figure 4.9 (A) General scheme of hydroamination of C_{60} fullerene with primary and secondary aliphatic amines. (B) Addition of morpholine to C_{60} fullerene in the presence of oxygen.

of various regioisomers, among which the 1,2- and 1,4-addition products are preferred (Figure 4.9A). A commonly accepted reaction mechanism for this type of reaction begins with an electron transfer from an amine to fullerene followed by either a radical coupling or proton transfer step. If oxygen is not excluded from the reaction system, it can participate in the reaction generating a complex mixture of products. For example, the reaction of C_{60} with an excess amount of morpholine in the presence of oxygen leads to several products, including a 1,4-bismorpholino[60]fullerene, a tetrakismorpholino[60]fullerene epoxide, and a morpholino[60]fullerene dimer (Figure 4.9B). Moreover, multiple additions of amines on C_{60} fullerene may readily occur. The number of amine additions is dependent on various factors, including the type of amine, the conditions of the experiment, and the solubility of the resulting products.

4.3.2 Cycloaddition

A well-established functionalization method for C_{60} fullerene is known as the **Bingel reaction** (Figure 4.10), in which a halo-substituted stabilized carbanion is first generated by treatment with a base and then undergoes an addition–elimination process to form a cyclopropanation product [14].

The Bingel reaction represents the most efficient method for the synthesis of methanofullerene derivatives. Besides the formation of mono-adducts of C_{60}, cycloaddition can occur multiple times on C_{60} under controlled conditions. Successive addition of independent reagents to C_{60} fullerene usually leads to a mixture of regioisomers and/or stereoisomers, which are difficult to separate. In order to achieve regio- or stereocontrol over the formation of multi-adducts of C_{60}, a tether-directed remote functionalization strategy was introduced by Diederich [15]. Figure 4.11 illustrates an example of a stereoselective synthesis of an enantiopure bisadduct of C_{60} fullerene through a double Bingel reaction directed by a chiral cleavable tether. In this synthesis, a tethered bis(malonate ester) is first converted into an active

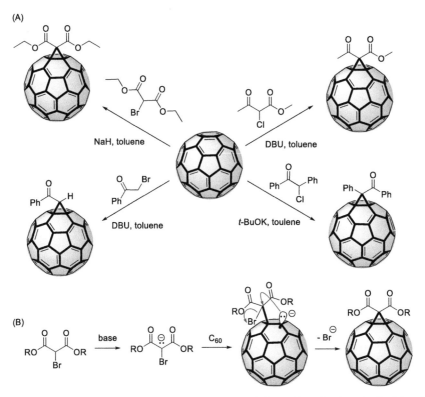

Figure 4.10 (A) Examples of the Bingel reaction. (B) General mechanism for the Bingel reaction.

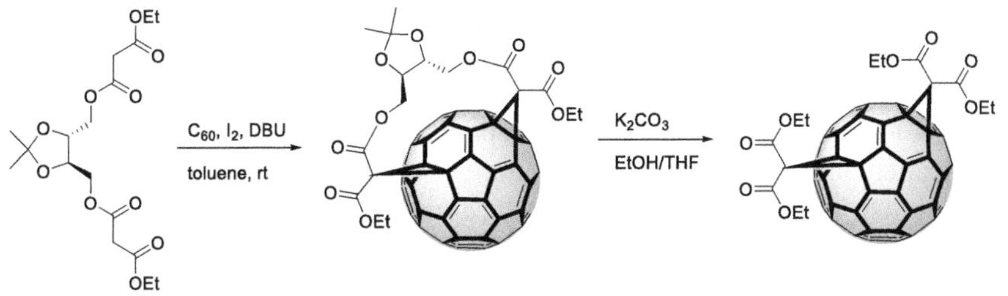

Figure 4.11 Stereoselective synthesis of a bis-adduct of C$_{60}$ fullerene through a tether-directed double Bingel reaction.

bis(iodo-malonate) *in situ* which then reacts with C$_{60}$ fullerene. The presence of the chiral tether group allows the enantiomeric bisadducts to be formed.

Cyclopropanation of C$_{60}$ fullerene can also be achieved through the addition of a diazo compound to form a pyrazoline intermediate, which undergoes N$_2$ extrusion

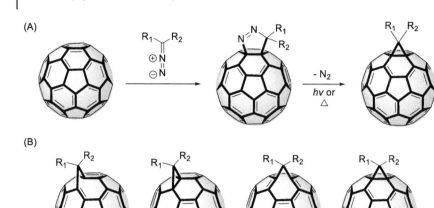

(A)

(B)

[5,6]-open   [5,6]-closed   [6,6]-open   [6,6]-closed

Figure 4.12 (A) Synthesis of methanofullerenes through the cycloaddition of diazocompound followed by N_2 extrusion. (B) Four possible isomeric structures of a methanofullerene.

through photolysis or under heating (see Figure 4.12A) [16]. The formation of a methanofullerene through this approach can lead to possibly four isomeric structures as depicted in Figure 4.12B, which are called [5,6]-open, [5,6]-closed, [6,6]-open, and [6,6]-closed, respectively. Experimentally, only [5,6]-open and [6,6]-closed products have been isolated and observed, while in most cases the [5,6]-open adducts are formed initially and then converted into thermodynamically more stable [6,6]-closed isomers.

Apart from cyclopropanation, [2+2], [3+2], and [4+2] cycloaddition reactions can also take place on the cage of C_{60} fullerene. For example, C_{60} fullerene can react with highly reactive benzyne, forming a four-member ring at the [6,6] bond (Figure 4.13A). Under photochemical conditions, C_{60} fullerene even reacts with itself through a photo [2+2] cycloaddition pathway to form a dimer (Figure 4.13B).

C_{60} fullerene is known to react with a range of 1,3-dipoles through [3+2] cycloaddition. The most notable example as well as one of the most popular method for functionalization of C_{60} fullerene is the **Prato reaction**, which uses an *in situ* generated azomethine yield (1,3-dipole) to react with a [6,6] bond of C_{60} fullerene (1,3-dipolarophile) to form a five-member pyrrolidine ring (Figure 4.14) [17]. Through the Prato reaction, various functional groups (R_1 and R_2) pre-installed in the synthetic precursors can be readily attached to C_{60} through robust covalent linkages. Like the Bingel reaction, the Prato reaction can also occur multiple times on a fullerene cage to afford fullerene multi-adducts. Moreover, the Prato reaction has been widely used to functionalize higher fullerenes as well as the sidewall of carbon nanotubes (CNTs) [18].

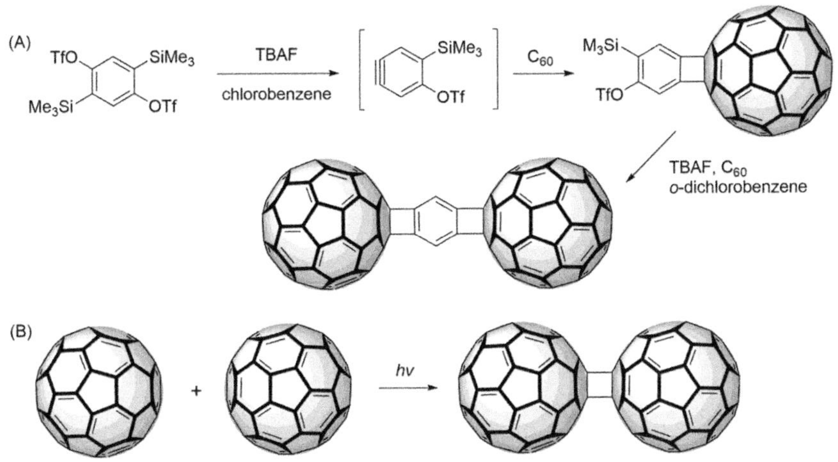

Figure 4.13 Exemplar [2+2] cycloaddition reactions that take place on the cage of C$_{60}$ fullerene.

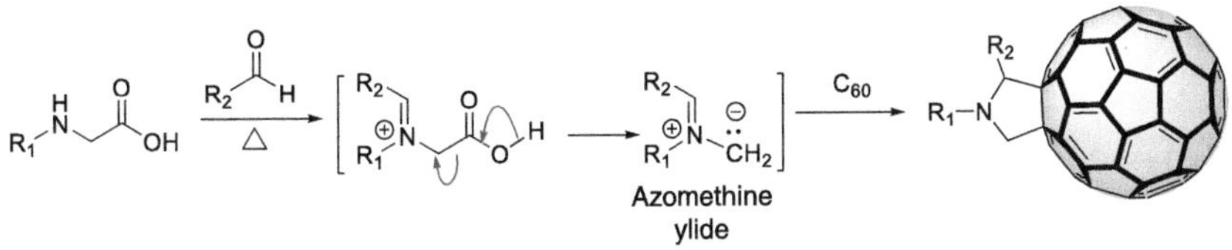

Figure 4.14 The Prato reaction on C$_{60}$ fullerene.

As discussed before, C$_{60}$ fullerene behaves as an electron-deficient olefin, which makes it a sufficient dienophile to participate in the Diels-Alder [4+2] cycloaddition [19]. C$_{60}$ fullerene is known to react with a large variety of dienes, such as 1,3-butadiene derivatives, activated furans, anthracene, cyclopentadienes, cyclo-heptatrienes, and *ortho*-quinodimethanes. In these reactions, the dienes are either directly applied to react with C$_{60}$ fullerene at elevated temperature (Figure 4.15A) or to be generate *in situ* and then react (Figure 4.15B). It is worth mentioning that the reversible nature of the Diels-Alder reaction makes it a useful blocking or directing strategy in the regioselective and stereoselective synthesis of various C$_{60}$ adducts.

4.3.3 Other Reactivities of C$_{60}$ Fullerene

In addition to the well-known reactions mentioned in the previous two subsections, C$_{60}$ fullerene can undergo a large variety of reactions, including radical addition, hydrogenation, oxidation, hydroxylation, Friedel-Crafts reactions, and

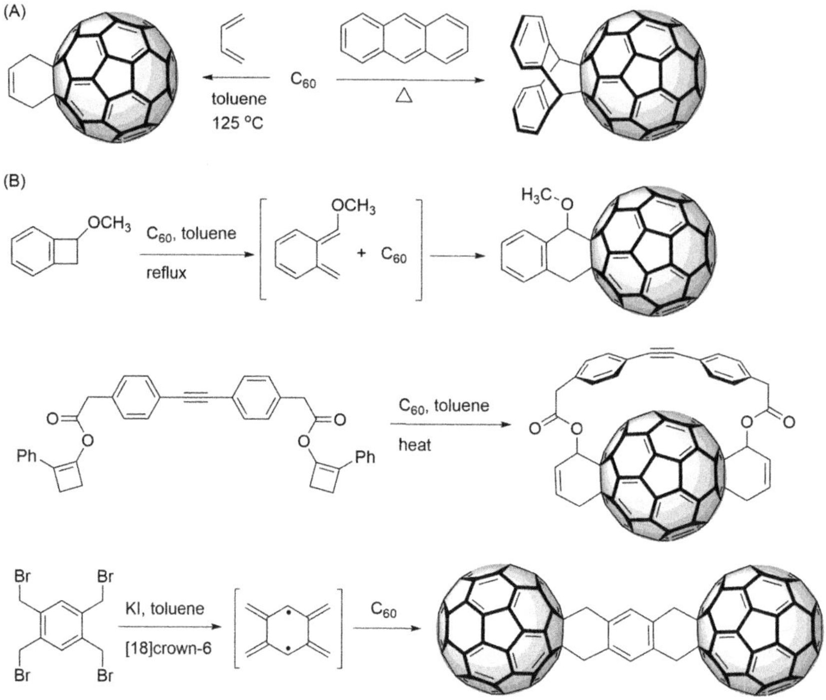

Figure 4.15 Examples of Diels-Alder [4+2] cycloaddition reactions on C_{60} fullerene.

just to mention a few. Many of these reactions, however, lead to the formation of mixtures of fullerene derivatives. To be synthetically useful, controlled experimental conditions are often needed. For example, free-radical reactions were one of the first types of reactions investigated in fullerene chemistry, but many of the early discovered radical reactions lacked selectivity due to the high affinity of C_{60} fullerene for radicals, thus limiting their application in the functionalization of C_{60}. Under appropriately controlled conditions, however, selective additions to C_{60} can be achieved. For example, by using tetrabutylammonium decatungstate (TBADT) as a radical initiator, various alcohol radicals can be generated under photochemical conditions, which then attack C_{60} fullerene to form mono-addition products (see Figure 4.16).

Some reactions can occur with C_{60} fullerene to open an orifice on its cage, leading to the so-called open-cage fullerenes. When the orifice is large enough, some molecules such as dihydrogen can be loaded into the inner cavity of C_{60} fullerene through the orifice created by the chemical modifications. An impressive example of this type of chemistry was demonstrated by Komatsu and co-workers in 2005 [20]. As illustrated in Figure 4.17, pristine C_{60} fullerene was subjected to a

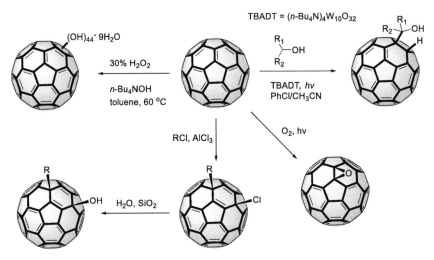

Figure 4.16 Various reactivities of C₆₀ fullerene.

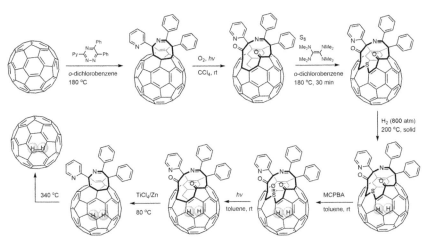

Figure 4.17 Synthesis of an endohedral fullerene, $H_2@C_{60}$, through a molecular surgical approach.

sequence of reactions to form a 13-membered ring orifice on its cage. This open-cage fullerene can encapsulate a dihydrogen molecule in a 100% yield to form an endohedral fullerene, which then undergoes four steps of organic reactions to completely close the 13-membered ring orifice. This molecular surgical method has a great application in making various **endohedral fullerenes**. Endohedral fullerenes (also called **endofullerenes**) are fullerenes that have additional atoms, ions, or small molecules enclosed within their inner spheres.

4.3.4 Organic Synthesis of C_{60} Fullerene

Although C_{60} fullerene can be produced on large scales using the resistive heating of graphite or combustion methods, the pursuit of its rational stepwise synthesis has garnered considerable interest from the synthetic community. An attractive synthetic approach is to construct small polyarenes that are the segments of C_{60} fullerene and then use certain bond forming reactions to "stitch" them together. However, this strategy faces the challenges of how to effectively introduce curvatures or pyramidalizations in the carbon network. In 1991, Scott and co-workers developed a method called flash vacuum pyrolysis (FVP), which can readily achieve the ring-closure to form the curved corannulene, the key segment of C_{60} fullerene [21]. As shown in Figure 4.18, the FVP process first generates terminal alkynes *in situ* through the thermal elimination of HCl, which subsequently cyclize to yield corannulene in a yield of 35–40%.

Employing the FVP as the key "stitching up" step, Scott and co-workers accomplished the first rational synthesis of C_{60} fullerene in 2002 [22]. As shown in Figure 4.19, the synthesis began with a commercially available chemical, 1-bromo-4-chlorobenzene. Through 12 steps of organic reactions, a truxene derivative ($C_{60}H_{27}Cl_3$) was prepared on a multi-gram scale. In the final synthetic step, the FVP reaction was applied to the precursor $C_{60}H_{27}Cl_3$ to generate C_{60} fullerene in an estimated yield of 0.1–1.0%. Although the overall yield of this synthesis is very low, the

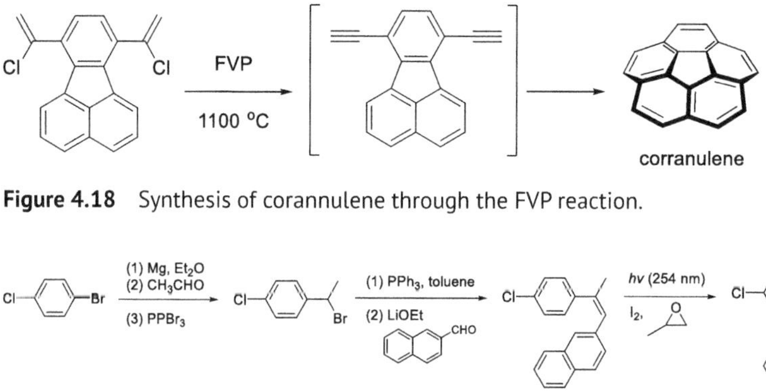

Figure 4.18 Synthesis of corannulene through the FVP reaction.

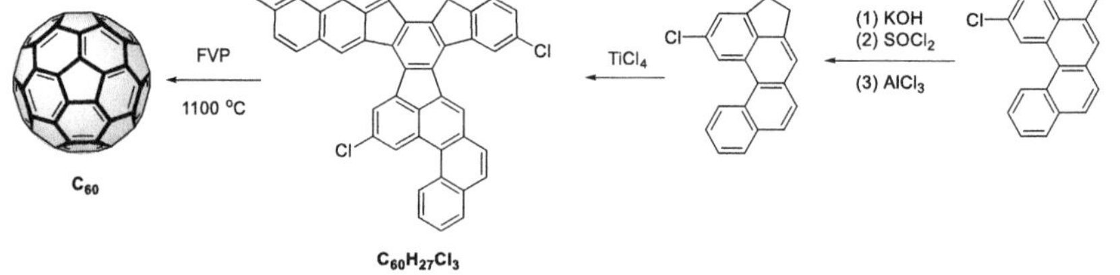

Figure 4.19 The first rational synthesis of C_{60} fullerene by Scott and co-workers.

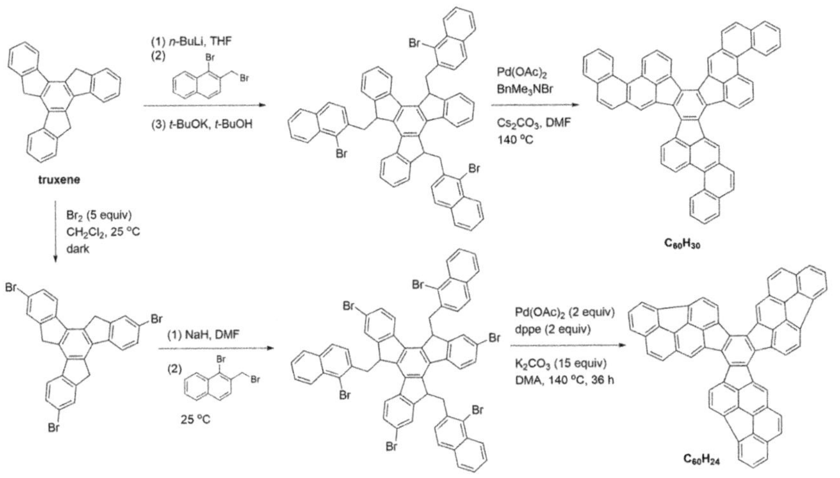

Figure 4.20 Multistep synthesis of crushed fullerene $C_{60}H_{30}$ and $C_{60}H_{24}$.

successful synthesis of C_{60} fullerene does provide fundamental insights into how to prepare fullerenes that are not accessible through the vaporization of graphite.

In addition to the FVP reaction, C_{60} fullerene can also be constructed through cyclodehydrogenation of C_{60} polyarenes, known as the "crushed fullerenes" under laser irradiation or upon heating on a platinum surface [23]. Figure 4.20 shows the synthesis of two crushed fullerenes $C_{60}H_{30}$ and $C_{60}H_{24}$, which can both lead to C_{60} fullerene when subjected to laser-induced cyclodehydrogenation reactions.

4.4 Chemistry of Carbon Nanotubes

Carbon nanotubes (CNTs) can be viewed as "cylindrical fullerenes," which possess diameters similar to those of fullerenes (from one to a few nanometers) but with much longer lengths. As mentioned earlier, carbon nanotubes can be divided into single-walled, double-walled, and multiple-walled carbon nanotubes. The preparation of CNTs can be achieved using various approaches, including the arc-charge method, laser ablation technique, and the catalytic chemical vapor deposition (CCVD) method. Among them, CCVD is so far the most widely used and investigated method owing to its remarkable ability to control the structures of CNTs formed, especially in the chirality-selective synthesis of SWCNTs.

4.4.1 Growth Mechanisms of Single-walled Carbon Nanotubes

The CCVD method for growing single-walled carbon nanotubes (SWCNTs) involves the catalytic decomposition of carbon sources (i.e., carbon-containing

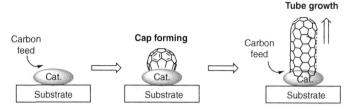

Figure 4.21 Scheme of the VLS mechanism for the growth of a SWCNT on a nanoparticle catalyst.

molecules) on nanoparticle (nanometer-sized) catalysts. The nanoparticles are typically those of various metals such as Fe, Cu, Co, Ni, Pt, Pd, Mn, Mo, Cr, Sn, Au, Mg, Al, and Rh. Non-metallic catalysts (e.g., SiC, Ge, Si, BN) and oxide catalysts (e.g., SiO_2, Al_2O_3, and TiO_2) have also been reported, but they are generally believed to be less efficient than the metallic catalysts for SWCNT growth. The nanoparticle catalysts can be either bound to a substrate or float at the tip of a growing carbon tube. The growth usually takes place at a high temperature (600–1100 °C) under complex conditions. There are several mechanisms to account for the growth of SWCNTs. Among them, the vapor–liquid–solid (VLS) mechanism depicted in Figure 4.21 is the one that is popularly used to explain the growth of SWCNTs on metallic nanoparticle catalysts. In this mechanism, carbon precursors first diffuse on the surface of a nanoparticle catalyst where they decompose into carbon species, which are dissolved to form a liquid alloy between metal and carbon. When the liquid catalyst particles are saturated with carbon, the carbon atoms begin to precipitate from the surface of each particle to form the cap of a SWCNT. As the process continues, a solid nanotube begins to extrude from the particle. The tube keeps growing on the catalyst particle until some event causes the growth to stop.

4.4.2 Chirality of SWCNT

SWCNTs show outstanding mechanical, electrical, and optical properties. As a result, SWCNTs have found a wide range of applications in science, engineering, biology, and medicine. As-prepared SWCNTs are not structurally homogeneous. Instead, they are mixtures of tubes with varied diameters, lengths, and chiral angles. Since a SWCNT can be conceptually viewed as being formed through rolling up a graphene sheet, the specific structure of the nanotube is actually defined by how the rolling is performed. As illustrated in Figure 4.22, the rolling of a graphene sheet can be carried out using different directions of axial wrap. Rolling along the symmetry axis results in the formation of zigzag or armchair nanotubes, the perimeter vectors of which lie along the directions of the two basis vectors $\mathbf{a_1}$ and $\mathbf{a_2}$ of the graphene lattice, respectively. Rolling along other angles

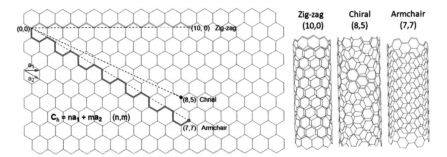

Figure 4.22 Schematic illustration of rolling up a graphene sheet to form SWCNTs with various chiral indices (n, m).

leads to the formation of chiral nanotubes with specific helicity. The exact structure of a carbon nanotube can be defined by the chiral index (n, m), which specifies its perimeter vector ($\mathbf{C_h}$, chiral vector) as shown in the following equation,

$$\mathbf{C_h} = n\mathbf{a_1} + m\mathbf{a_2} \tag{4.1}$$

where n and m are integers that specify the diameter and helicity of the tube. Zigzag and armchair SWCNTs are achiral, their chiral indices being $m = 0$ and $n = m$, respectively. Other combination of n and m gives chiral nanotubes. The electronic properties of SWCNTs are significantly dependent on their chiral indices. When $|n - m| = 3q$ (q is an integer), the nanotubes show metallic or semi-metallic behavior. In all other cases, the nanotubes are semiconducting in electronic nature. When the n and m values are known, the diameter of a SWCNT (d) can be calculated using the following equation,

$$d = \frac{a_0}{\pi} \sqrt{n^2 + m^2 + nm} \tag{4.2}$$

where $a_0 = 0.246$ nm. Moreover, the angle between the perimeter vector and the basis vector $\mathbf{a_1}$ is called the helical angle (α), which is defined by Eq. 4.3. As can be seen from Figure 4.22, the zigzag nanotubes have a helical angle of $\alpha = 0°$, while the armchair nanotubes have an $\alpha = 30°$.

$$\alpha = tan^{-1} \left(\frac{\sqrt{3}m}{2n + m} \right) \tag{4.3}$$

4.4.3 Non-Covalent Functionalization of SWCNTs

The production of CNTs, particularly SWCNTs, provides a mixture of tubes with different lengths, diameters, and chiral indices. The electronic and optical properties of various types of SWCNTs are very different. For applications in advanced electronics and optoelectronics, the electronic nature and structural homogeneity of CNTs are required knowledge. It is therefore important to have the nanotube mixtures differentiated according to their structural and electronic properties (e.g., semiconducting, metallic). Unlike fullerenes which can be solubilized in certain organic solvents, SWCNTs show a strong tendency to form bundles and aggregates, hence they are insoluble in most solvent media. Structural heterogeneity and bundling issues are two major challenges that hinder the applications of SWCNTs in advanced electronic and optoelectronic devices. Effective functionalization of SWCNTs has been investigated as a solution to these problems.

Functionalization of SWCNTs can be achieved by two different approaches, namely non-covalent and covalent functionalization. The non-covalent functionalization of SWCNTs is generally achieved by applying dispersants to disrupt the nanotube bundles through non-covalent interactions. The nanotube-dispersant interactions result in individualized nanotubes that exist as a stable suspension in a solvent. Commonly used dispersants for SWCNTs range from various surfactants, synthetic and natural polymers, to π-conjugated oligomers and DNA molecules. Surfactants such as sodium dodecylbenzene sulfonate (SDBS) can suspend SWCNTs in water at relatively high concentrations (up to ~20 mg/mL). Usually, the dispersion processes are done by ultrasonicating SWCNTs in a surfactant solution that exceeds the critical micelle concentration (CMC) of the surfactant. Figure 4.23 schematically depicts the process of dispersion of SWCNTs with a surfactant dispersant. Depending on the experimental conditions, the surfactant molecules may form cylindrical micelles, semi-micelles, or are randomly adsorbed on the surface of the dispersed nanotubes through van der Waals forces, π-π interactions, and hydrophobicity/hydrophilicity interactions.

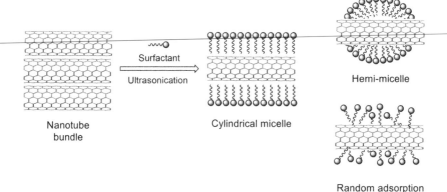

Figure 4.23 Schematic illustration of surfactant-induced dispersion of SWCNTs and three possible interaction modes between SWCNTs and surfactant molecules.

Once dispersed as individual nanotubes in solution, SWCNTs can be further sorted out according to length, diameter, chirality, and electronic nature (metallic versus semi-conducting) by utilizing different strategies. Taking advantage of differences in surfactant organization on suspended individual nanotubes, a method based on the **density gradient ultracentrifugation** (DGU) technique has been developed and successfully used to achieve separation of SWCNTs. DGU is a method widely used in biochemistry for the purification of proteins and nucleic acids. In the sorting of SWCNTs, an ultracentrifugation process is applied to the suspension of surfactant-encapsulated SWCNTs, driving them to migrate into the density gradient medium until they reach their corresponding **isopycnic points** (i.e., the points where their buoyant density equals that of the surrounding medium). Actually, this method is so effective that it can successfully achieve separation of enantiomers of chiral nanotubes. Figure 4.24 shows the separation of two types of enantiomerically pure (6,5) SWCNTs as demonstrated by Weisman and co-workers in 2010 [24]. In this work, SWCNTs were dispersed in a mixture of water and iodixanol in the presence of sodium cholate (surfactant) with the aid of ultrasonication. The suspension was carefully subjected to nonlinear density gradients created by the DGU technique to achieve advanced separation of enantiomerically pure nanotubes.

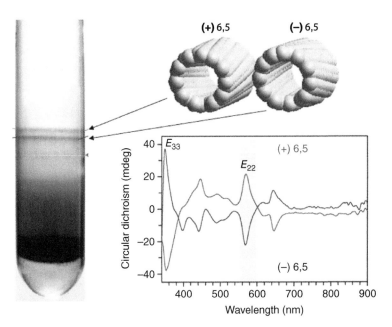

Figure 4.24 Photographic image of a centrifuge tube with distinct bands of enantiomers of (6,5) SWCNTs (left) and the circular dichroism (CD) spectra of the separated enantiomers. (Reproduced from *Nat. Nanotechnol.* 2010 /Ghosh et.al.,2010 / Springer Nature).

Single-strain DNA (ss-DNA) can be adsorbed on the surface of SWCNTs to form water soluble DNA-nanotube hybrids. The effective net charge of the hybrids is then dependent on how ss-DNA is arranged on the nanotubes as well as the electronic properties of the nanotube. The wrapping of a negatively charged DNA backbone around a metallic nanotube would result in a lowered net linear charge, but if the metallic nanotube is replaced with a semiconducting nanotube, the ss-DNA wrapping causes an increased effective linear charge. Differences in net linear charge allow the ss-DNA/nanotube solutions to be fractionated by methods such as ion-exchange chromatography [25]. Molecular dynamics (MD) simulation studies (Figure 4.25) have indicated that the DNA/nanotube interactions are dominantly controlled by the π-π stacking of the DNA base groups and the aromatic surface of the nanotube as well as the conformation of the DNA backbone [26].

π-Conjugated polymers show strong interactions with the sidewall of SWCNTs through π-π interactions, and therefore have been extensively explored as dispersants for carbon nanotubes. Certain linearly π-conjugated polymers, such as poly(phenylene vinylenes) (PPV), poly(*p*-phenylene ethynylene)s (PPE), poly (phenylene butadiynylene)s (PPB), and poly(thiophene)s (PT), have been found to show strong affinity for the sidewall of SWCNTs (see Figure 4.26). The complexation

Figure 4.25 Model of a DNA sheet wrapping around a (8,4) nanotube. (Reproduced from Nano Lett. 2009 / Johnson et.al.,2009 / American Chemical Society).

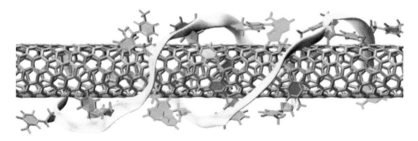

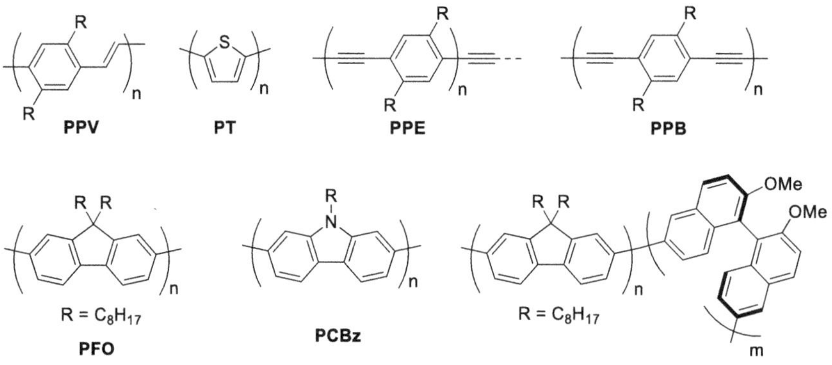

Figure 4.26 Selected examples of π-conjugated polymers commonly used as dispersants for SWCNTs.

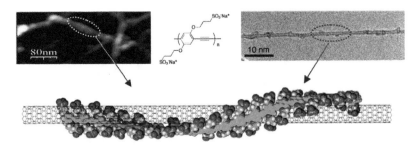

Figure 4.27 AFM (top left) and TEM (top right) images of individual SWCNTs wrapped by an amphiphilic PPE. Snapshot of the interactions of a (10,0) tube with a PPE 20-mer obtained through MD simulations at 20 ns. (Reproduced from Nano Lett. 2009 / Kang et.al.,2009 / American Chemical Society).

of these polymers with CNTs leads to polymer-nanotube hybrids that can be well dispersed as stable individual nanotube hybrids in solution. Figure 4.27 illustrates the dispersion outcomes of SWCNTs with an amphiphilic linear PPE in water [27]. Based on atomic force microscopic (AFM) and transmission electron microscopic (TEM) analyses in combination with molecular dynamics (MD) simulations, the PPE molecules have been proven to helically wrap around the backbone of single-walled carbon nanotubes through π-π stacking, resulting in water-soluble hybrids.

A vast array of π-conjugated polymers has been developed and investigated for non-covalent functionalization of SWCNTs with the aim of selectively sorting out specific types of tubes. In this respect, fluorene, thiophene, and carbazole-based conjugated polymers and co-polymers have received enormous attention. It has been demonstrated that the selectivity of a certain conjugated polymer for dispersing SWCNTs is dependent on numerous factors, including the electronic and conformational properties of the polymer, the type of raw CNTs, sonication time/temperature, and solvent. Usually, the π-conjugated polymers show strong affinity for SWCNTs and therefore are irreversibly adsorbed on the sidewall of nanotubes. To enable the polymer-nanotube assemblies to be dissociated, some reversible strategies have been designed based on stimuli-responsive conjugated polymers. Figure 4.28 illustrates an example of the reversible dispersion of SWCNTs using a rationally designed redox-switchable conjugated polymer. In this work, Adronov and co-workers prepared a conjugated co-polymer that contains a tetrathiafulvalene vinylogue (TTFV) and two fluorene groups in its repeat unit (see Figure 4.28B) [28]. The TTFV group acts as a redox-active unit that exhibits redox-controlled conformational switching behavior. Structurally, TTFV is a π-extended analogue of the well-known organic π-donor, tetrathiafulvalene (TTF). In its neutral state, the molecule adopts a *cis*-like conformation (Figure 4.28A). Upon oxidation or protonation, TTFV can be converted into a dication, in which the two positively charged dithiolium rings prefer to take a *trans* conformation as a result of electrostatic

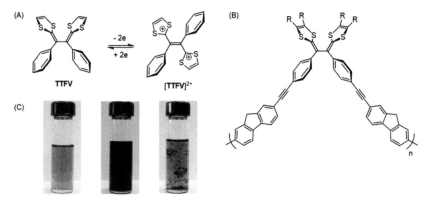

Figure 4.28 (A) Redox-controlled conformational switching properties of TTFV. (B) Structure of a TTFV-fluorene co-polymer designed as a stimuli-responsive dispersant for SWCNTs. (C) Photographical images of TTFV-fluorene co-polymer in toluene (left), SWCNTs dispersed in the polymer solution (middle), and the polymer solution with SWCNTs after addition with TFA (right). (Reproduced from J. Am. Chem. Soc. 2014 / Liang et.al.,2014 / American Chemical Society).

repulsion. The TTFV-fluorene co-polymer in the neutral state takes a helical conformation, allowing it to efficiently wrap around individual SWCNTs to form stable polymer-nanotube assemblies that are soluble in organic solvents (Figure 4.28C, middle). Upon addition of a strong organic acid, trifluoroacetic acid (TFA), into the dispersion, the TTFV units in the co-polymer are ionized to give conformational changes. As such, the co-polymer unwraps itself from the SWCNTs, leaving the protonated co-polymer soluble in the solvent but the released SWCNTs as a precipitate. This stimuli-responsive polymer approach not only allows a specific subpopulation of SWCNTs from an impure mixture to be selectively dispersed in organic solvents, but can conveniently release and isolate the dispersed nanotubes in pure form by application of a simple release trigger (e.g., acidification).

SWCNTs can also interact with polycyclic aromatic hydrocarbons (PAHs) that are properly functionalized with solubilizing groups through π–π stacking to form stable dispersion in aqueous or organic media. Among various PAHs, pyrene has been widely used as an anchoring group for binding with the graphitic surface of SWCNTs. Figure 4.29 shows an elegant example reported by Dai and co-workers in 2001, in which a 1-pyrenebutanoic acid succinimidyl ester is first irreversibly adsorbed on the surface of SWCNTs through π–π stacking [29]. The N-hydroxysuccinimidyl (NHS) ester group attached to the pyrene is highly reactive toward nucleophilic acyl substitution by a primary or secondary amine group under physiological conditions. Such amino groups exist in abundance on the surface of most proteins. Taking advantage of this reactivity, proteins or small

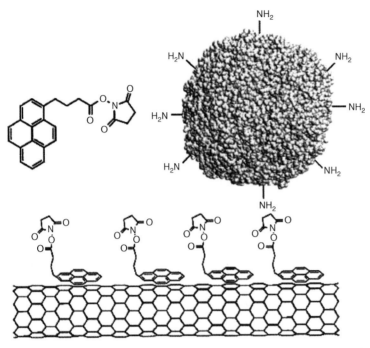

Figure 4.29 Immobilization of proteins on the sidewall of SWCNTs through 1-pyrenebutanoic acid succinimidyl ester. (Reproduced from *J. Am. Chem. Soc.* 2001, *123*, 3838–3839 with modification).

biomolecules can be further immobilized on the surface of such pyrene-functionalized SWCNTs.

4.4.4 Covalent Functionalization of SWCNTs

SWCNTs can be covalently functionalized through numerous organic reactions to yield functionalized CNTs. Owing to the curved structure of SWCNTs, the sp^2 hybridized carbons are pyramidalized and the π-orbitals between adjacent carbon atoms are misaligned. As a result, all of the sp^2 carbons carry a significant amount of strain energy, making them more reactive than those on a planar graphene sheet. In a SWCNT, the endcap unit resembles a hemispherical fullerene and the carbon atoms in this region are therefore more reactive than those on the sidewall. SWCNTs can undergo various chemical reactions, with the relief of the carbon strain energy being the main driving force. It is worth noting that covalent reactions on SWCNTs invariably produce sp^3 hybridized carbons, which in turn modify the electronic properties of the SWCNTs. It is therefore important to consider the balance of the degree of functionalization with the resultant changes in electronic properties.

A commonly used method for non-covalent functionalization of SWCNTs is oxidation. When CNTs are treated with strong oxidizing agents, such as nitric acid or sulfuric acid or a mixture of both under ultrasonication, the ends and sidewall of the nanotubes can be functionalized with carboxylic acid and/or quinone groups, usually at the defect sites. With the carboxyl groups covalently attached, nanotubes can be further modified with diverse molecular or biomolecular functionalities through efficient esterification or amine-carboxylic acid coupling reactions (Figure 4.30). These types of functionalizations have been widely used in making water-soluble SWCNTs for biological applications. SWCNTs can be subjected to reductive alkylation under Birch reduction conditions. A well-established method is known as the **Billups protocol** (Figure 4.30), which uses lithium and alkyl halides to functionalize SWCNTs in liquid ammonia [30]. This method is scalable and can readily achieve a high degree of functionalization to prepare soluble functionalized CNTs.

The C–C bonds on the sidewall of a SWCNT show double bond character and hence can undergo cycloaddition reactions similar to those occurring on a fullerene cage. For example, an alkyl azide can be decomposed to yield a reactive nitrene species under thermal or photochemical conditions. The resulting nitrene shows a preference for attacking the sidewall of a SWCNT through a [2+1] cycloaddition pathway (Figure 4.31). Additionally, the Bingel and Prato reactions can also occur on the sidewall of SWCNTs, serving as efficient methodologies for covalent linking of functional groups to SWCNTs.

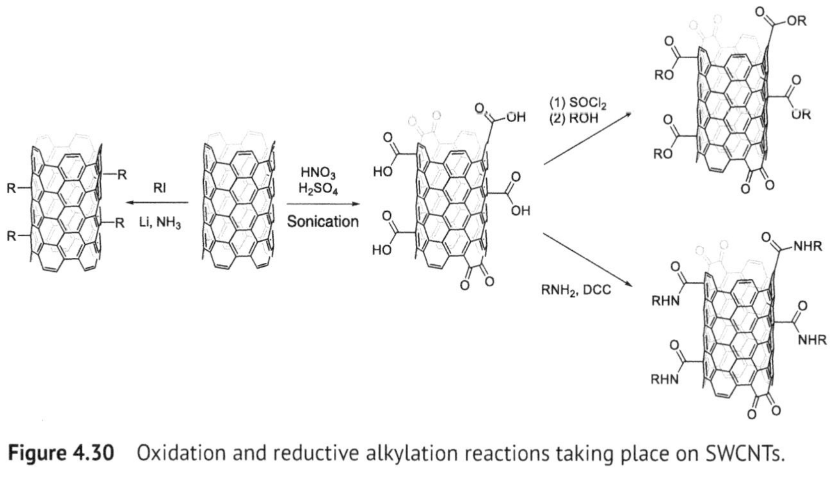

Figure 4.30 Oxidation and reductive alkylation reactions taking place on SWCNTs.

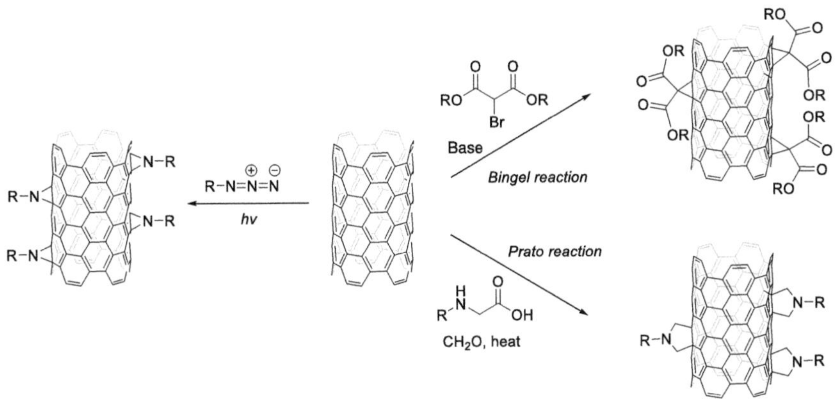

Figure 4.31 Covalent functionalization of SWCNTs through cycloaddition reactions.

The sidewall of SWCNTs is also susceptible to other highly reactive species such as fluorine and organic radicals. It has been found that at a temperature higher than 150 °C fluorine can significantly react with SWCNTs to form fluorinated products [31]. The fluorinated nanotubes can be de-fluorinated by treatment with hydrazine or through reaction with alkylmagnesium bromides or alkyllithium reagents to yield alkylated SWCNTs [32].

As introduced in Section 2.4 of Chapter 2, arene diazonium salts can generate aryl radicals at elevated temperatures, showing a high degree of reactivity toward various carbon surfaces including the sidewall of SWCNTs. In 2003, Strano and co-workers discovered that water-soluble arene diazonium salts are able to extract electrons from carbon nanotubes in the process of forming a covalent bond with CNTs [33]. Such reactions show high chemoselectivity for metallic tubes in the presence of semiconducting tubes. As illustrated in Figure 4.32A and 4.32B, the C–C bond between an aryl and the surface of a nanotube is formed with extremely high affinity for electrons with energies (ΔE_r) near the Fermi level (E_f) of the nanotube. **Fermi level** is a concept in solid-state physics. For semiconducting materials, Fermi level represents the highest energy level occupied by electrons in the ground state system at 0 K (also known as Fermi energy), which is an important parameter describing the electrical characteristics of a solid material. The arene diazonium group forms a charge-transfer complex at the nanotube surface, where electron donation from the latter stabilizes the transition state and accelerates the forward rate of reaction. Once the bond symmetry of the nanotube is disrupted by the formation of this defect, adjacent carbons gain increased reactivity (Figure 4.32C), and the initial selectivity is amplified as the entire nanotube is functionalized. As such, metallic carbon nanotubes exclusively react with the diazonium salt under controlled conditions. It is remarkable to note that the functionalized carbon nanotubes obtained can be then converted back

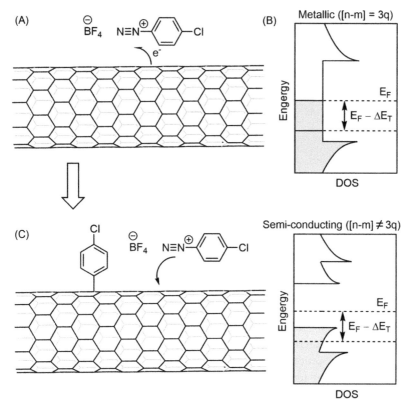

Figure 4.32 (A) Diazonium reagents extract electrons, thereby evolving N₂ gas and leaving a stable C–C covalent bond with the nanotube surface. (B) The extent of electron transfer is dependent on the density of states (DOS) in that electron density near E_f leads to higher initial activity for metallic and semi-metallic nanotubes. (C) The arene-functionalized nanotube may now exist as a delocalized radical cation, which could further receive electrons from neighboring nanotubes or react with fluoride or diazonium salts.

to unfunctionalized nanotubes through thermal treatment, restoring the pristine electronic structure of the nanotube. In this way, metallic and semiconducting SWCNTs can be separated with high selectivity and scalability.

4.5 Chemistry of Graphene, Graphene Oxide, and Reduced Graphene Oxide

Graphene is a 2D carbon allotrope that ideally should take a fully planar structure with infinite dimensions. Real graphene, however, shows edges that are either in a zigzag or an armchair arrangement. It is interesting to mention that graphene with jagged edges can readily pierce cell membranes, enabling graphene to enter the cell to influence cell function. Graphene can be viewed as a macromolecule with a molecular weight greater than 10^6 g mol^{-1}. High-resolution microscopic analyses have revealed that a graphene surface often contains holes and defects, which cause locally curved domains and doping effects to influence electronic properties and reactivity. The extended planar π–surface of graphene enables it to show a pronounced propensity for π–π stacking, leading to the thermodynamically stable carbon allotrope graphite. A solid sample of graphene cannot stably exist unless it is on a surface to support it. Chemical functionalization, both covalent and non-covalent, can somewhat "mask" the surface of a graphene sheet, enabling it to stably exist as an individual entity without aggregation or stacking [34]. It is also worth mentioning that two π–π stacked graphene layers are often called bilayer graphene. A stack of graphene containing less than ten layers is called few-layer graphene. Mono-, bi- and few-layer graphene can be obtained from exfoliation of high-quality bulk graphite through mechanical shearing in a liquid or acid oxidation of graphite. The acid oxidation method is a well-developed and scalable method for generating graphene through wet chemistry. For example, slow addition of potassium permanganate (KMnO$_4$) to a solution of sodium nitrate (NaNO$_2$) and powdered flakes of graphite can effectively exfoliate graphite into graphitic oxide. This method was first reported by Hummers and co-workers in 1958 and now is well known as the **Hummers method** [35]. Graphene resulting from acid oxidation treatment usually contain various oxygenated groups (e.g., COOH, OH, and epoxy) and is therefore referred to as **graphene oxide** (GO). These oxygen groups generate electrostatic repulsion to facilitate the solvation of GO. On the other hand, the presence of oxygen groups on the basal plane of graphene disrupts the π-conjugation, making GO an insulating material with a larger optical band gap than graphene. Reductive treatment of GO can lead to the formation of **reduced graphene oxide** (rGO), which may yield tunable fluorescence in the near infrared (NIR) and infrared (IR) spectral range. Such properties are beneficial for biological applications, since cells and tissues usually do not exhibit auto-fluorescence in this region.

GO and rGO are suitable for further chemical modifications through reactions commonly applied to other types of polymers and macromolecules. Figure 4.33 illustrates a few examples of covalent functionalization of GO and rGO. For instance, an arenediazonium salt can react with rGO to have both sides of the graphene surface functionalized (Figure 4.33A). The mechanism for this reaction is commonly believed to consist of two key steps. First, a single electron transfer (SET) from the graphene to the arenediazonium cation induces the formation of a highly reactive aryl radical along with dinitrogen (N$_2$) gas. Second, the aryl radical attacks a carbon atom of the graphene sheet to form a C–C bond that links the aryl group to the graphene surface. The epoxy rings in GO can be subjected to ring opening reactions using various nucleophiles such as amines, alcohols, and azide ion. The hydroxy groups can be utilized as the synthetic handles for linking other

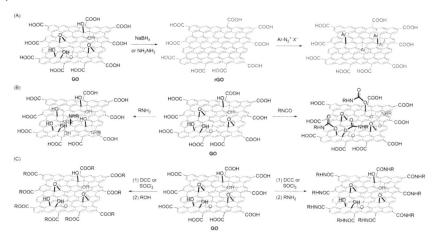

Figure 4.33 Various synthetic routes to covalently functionalize GO taking advantage of the oxygen groups present in its structure.

groups to graphene through ester or ether linkages (Figure 4.33B). The most common types of chemical reactions used to functionalize GO take advantage of the carboxyl groups present in its structure. Well-established synthetic methods such as carbodiimide or thionyl chloride activation allow the carboxyl groups to be efficiently converted into amides or esters (Figure 4.33C).

Although the chemistry of GO and rGO has been developed on an ongoing basis, the functionalization of GO and rGO still faces significant challenges. GO is a non-stoichiometric material, since the density of oxygen groups on GO cannot be precisely determined or controlled. Such structural heterogeneity inevitably leads to non-stoichiometric functionalization after GO is subjected to various chemical reactions. For electronic applications, GO is not an ideal material due to its non-conductivity as a result of the disruption of π-conjugation by the oxygenated groups attached to its surface. The oxygen content of GO can be considerably lowered through chemical reduction or thermal treatment, giving more stable rGO with increased conductivity. The defects in rGO are chemically activated, allowing them to be further functionalized for various applications, such as the production of transparent conductors and biosensors. The functionalization of the basal plane of graphene is also very challenging due to the relatively low reactivity of the planar sp^2 carbon atoms. The use of reactive species such as organic radicals can somewhat overcome this barrier, but the reactions are often plagued by unwanted side reactions and ill-defined structures. More controlled and selective reactions can be used to overcome this hurdle. Figure 4.34 illustrates an efficient functionalization approach for graphene sheets using azidated graphene as the intermediate [36]. In this work, graphene sheets were first grown on the surface of a SiO_2/Si substrate through the chemical vapor deposition (CVD)

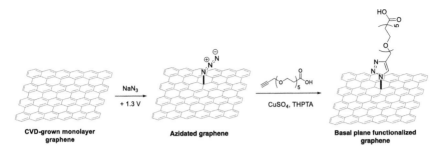

Figure 4.34 Functionalization of graphene sheet through azidation followed by click chemistry.

method. The CVD method is a "bottom-up" approach in which carbon-containing molecules are thermally decomposed on a metal surface to form graphene sheets. It has become the most highly used method for the preparation of large-area graphene on metallic surfaces. The CVD-grown monolayer graphene was then reacted with sodium azide (NaN_3) to yield azidated graphene. The azidation reaction was found to be more efficient when a positive voltage was applied to the graphene. With the azido groups attached to the surface of graphene, highly efficient click reactions such as CuAAC can be performed, allowing various functional groups to be covalently linked to the basal plane of graphene.

Graphene and rGO can be non-covalently functionalized using various molecular and macromolecular systems. In particular, π-conjugated polycyclic aromatic hydrocarbons and related polymers (e.g., pyrene, porphyrin, perylene diimide, and DNA) have been found to exhibit attractive π–π stacking and C–H$\cdots\pi$ interactions with the hydrophobic basal plane of graphene or rGO. Figure 4.35 shows an example of non-covalent functionalization of rGO with a [60]fullerene derivative containing a pyrene tag [37]. The pyrene group shows strong π–π stacking with rGO, allowing the [60]fullerene derivative to be immobilized on its surface. This rGO-pyrene-fullerene hybrid material was successfully applied as an electron extraction layer for bulk heterojunction (BHJ) polymer solar cells to give enhanced solar cell performance.

Besides π–π and C–H$\cdots\pi$ interactions, electron donor–acceptor complexation, hydrogen bonding interactions, and van der Waals forces can also be utilized as the driving forces for non-covalent functionalization of graphene, GO, and rGO. Compared with covalent functionalization, the non-covalent functionalization approach is largely preferred when the electronic properties of graphene need to be preserved. Moreover, non-covalent functionalization can flexibly introduce various new functionalities onto the surface of graphene, leading to enhanced dispersibility, biocompatibility, reactivity, binding capacity, or device performance (e.g., sensing function).

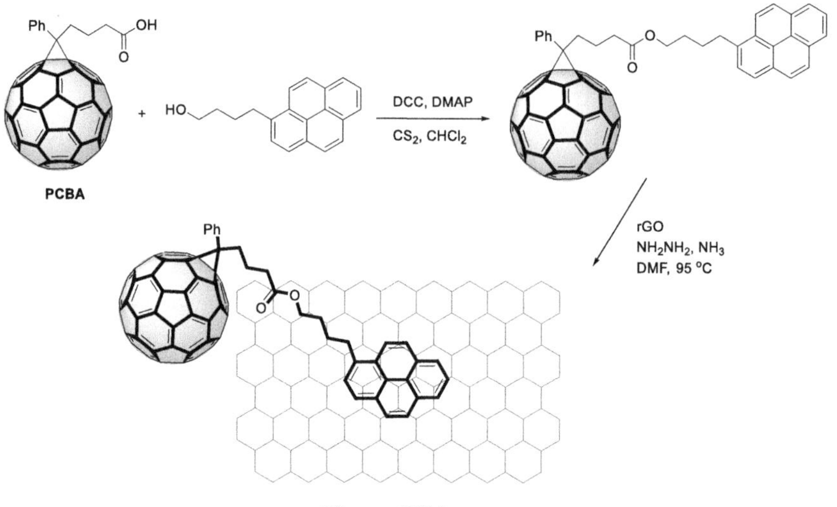

Figure 4.35 Non-covalent functionalization of rGO with a pyrene-[60]fullerene derivative.

4.6 Chemistry of Carbon Quantum Dots

In an effort to separate carbon nanotubes using electrophoresis, Scrivens and co-workers in 2004 surprisingly discovered a new class of highly fluorescent nanocarbon materials, which were later termed **carbon quantum dots** (CQDs) [38]. CQDs have less defined structures but show highly intriguing electronic and fluorescence properties. Spectroscopic studies have revealed that the structure of CQD is made of both graphene-type sp^2 hybridized and a relatively high amount of the diamond-type sp^3 hybridized carbon atoms. Therefore, CQDs are quasi-spherical nanoparticles with amorphous to nanocrystalline cores, in which graphene, GO sheets, and diamond-like structures are fused together. There are also various oxygenated groups (e.g., COOH, C=O, OH) present on the surface of CQDs (see Figure 4.36). The variations of these surface groups depend on the synthetic conditions.

The defects (or disorders) on the surface of CQDs are the main reason for why they show fluorescence emission. Usually, the CQDs emit in the visible region of the electromagnetic spectrum from blue to green color. Fluorescence quantum yields of CQDs have been reported at the relatively high level of 30–40%. Synthetic approaches for CQDs are generally classified into two categories, top-down and bottom-up. The top-down approach involves breaking down relatively large carbon sources, ranging from nanodiamonds graphite, CNTs, to carbon soot, activated carbon, and graphite oxide, by methods such as arc discharge, laser ablation, and electrochemical oxidation. The bottom-up approach generates CQDs from molecular precursors such

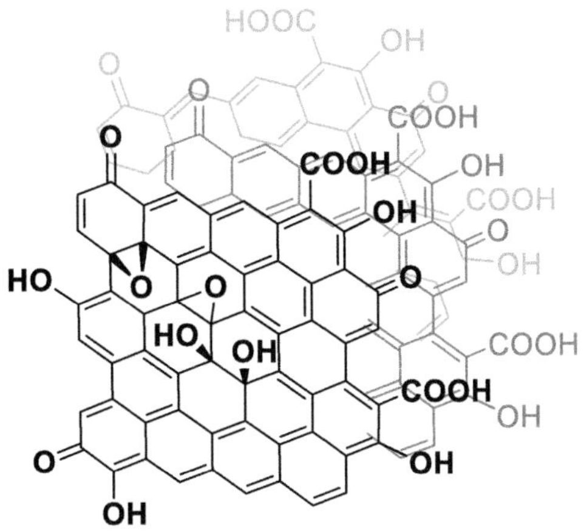

Figure 4.36 Proposed chemical structure of CQDs.

as citrate, carbohydrates, and polymer-silica nanocomposites through combustion/thermal treatment and supported synthetic and microwave synthetic routes. The surface of CQDs is highly sensitive to contaminants in their environment. To avoid the influence of various contaminants, surface passivation of CQDs is performed typically through coating a thin insulating layer of polymeric materials (e.g., polyethylene glycol) on an acid-treated CQD surface. Effective surface passivation has been found to be an essential step for CQDs to exhibit high fluorescence intensities. Functionalization of CQDs can introduce functional groups to yield different defects on the CQD surface. These defects work as traps for excitation energy and hence lead to significantly modified fluorescence emissions. CQDs have found extensive applications in fluorescence sensing, bioimaging, nanomedicine, light-emitting devices, electrocatalysis, and photocatalysis [39].

4.7 Chemistry of Molecular Nanocarbons

The discovery of various carbon nanomaterials (e.g., fullerenes, carbon nanotubes, graphene, carbon quantum dots, etc.) has considerably promoted the advancement of modern materials science and technology. Parallel to these developments, the synthetic pursuit of molecular analogues of large nanocarbon structures has been long capturing the attention of the organic synthetic community. These materials are dubbed **molecular nanocarbons** as they possess precise molecular structures which represent or resemble the segments of the fascinating carbon nanomaterials such as fullerenes, carbon nanotubes, and graphene. The following section will discuss a range of intriguing molecular nanocarbons according to their topological and chemical features.

4.7.1 Molecular Bowls

As discussed previously, the endcap segment of C_{60} fullerene or a closed-end CNT is a π-conjugated structure containing sp^2 carbon atoms and taking a hemispherical (or bowl) shape. The following are two molecules that closely represent such a unique molecular motif, namely corannulene and sumanene (Figure 4.37).

Corannulene is also known as dibenzo[*ghi,mno*]fluoranthene or [5]circulene. The name **corannulene** derived from Latin, having *annuli* within an *annulus* structure. The name of corannulene also refers to its structural relation with coronene. Corannulene has a structure containing a central pentagon fused with six hexagons. The central pentagon brings about considerable strain. To relieve this energy, the molecule takes a curved and bowl molecular shape with a C_{5v} symmetry. Corannulene has been estimated to have a strain energy of 24.2 kcal mol^{-1} [40]. Crystallographic studies have shown that the bowl depth of corannulene is 0.87 Å and it can undergo a bowl-to-bowl inversion with a barrier of 10.2 to 11.5 kcal mol^{-1} [41]. Theoretical studies have indicated that the central five-membered ring of corannulene is antiaromatic, while the outer six-membered rings are aromatic. Corannulene also serves as a proper model for understanding the magnetic properties of fullerenes. It has been reported that the central pentagon of corannulene shows a paratropic current, while the rim is diatropic.

The first synthesis of corannulene was achieved by Barth and Lawton in 1971, who performed a total of 17 linear synthetic steps to prepare this intriguing molecule in an overall yield less than 1% [42]. Ever since then, several synthetic approaches have been established to prepare corannulene with improved yields. For example, the FVP synthetic approach for corannulene devised by Scott (see Figure 4.18). The most efficient synthetic approach known so far is the one developed by Siegel and co-workers in 2011 (Figure 4.38), which comprised 9 steps and allowed for kilogram-scale production of corannulene [43]. Siegel's method has allowed corannulene to become a commercially available compound. Extensive

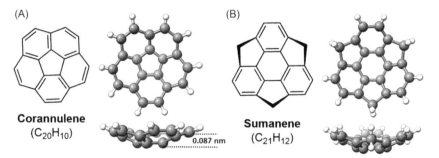

(A)

Corannulene
($C_{20}H_{10}$)

0.087 nm

(B)

Sumanene
($C_{21}H_{12}$)

Figure 4.37 Chemical structures and single crystal X-ray structures of (A) corannulene and (B) sumanene.

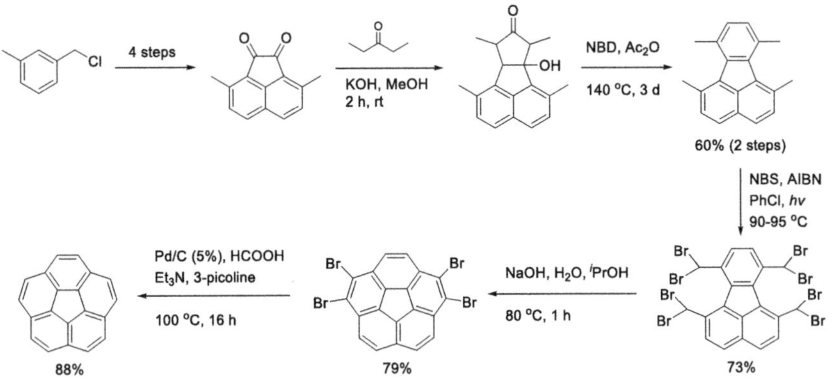

Figure 4.38 Siegel's kilogram-scale production of corannulene.

studies of its fundamental properties and applications have shown tremendous development since Siegel and co-workers original publication.

Sumanene is a C_{3v} symmetric polycyclic aromatic hydrocarbon which consists of three benzene and three cyclopentadiene rings alternatively fused together to form a central six-membered ring (Figure 4.37B). Like corannulene, the conformation of sumanene takes a bowl shape as a result of the strain within the molecule. The term **sumanene** was coined by Prof. Goverdhan Mehta in a communication article published in 1993 based on an attempted synthesis of sumanene. Herein, "suman" means flower in Hindi and Sanskrit to indicate the resemblance of the molecular shape to a flower [44]. In 2003, Sakurai and co-workers successfully achieved the first synthesis of sumanene through the synthetic route illustrated in Figure 4.39 [45]. In their synthesis, the *syn*-product of the trimerization of norbornadiene was subjected to tandem ring-opening metathesis (ROM) and ring-closing metathesis (RCM) reactions in the presence of the first-generation Grubbs catalyst. This reaction rapidly constructed the framework of sumanene. The final step was a dichlorodicyanoquinone (DDQ) promoted dehydrogenative aromatization that afforded sumanene in a good yield. Theoretical analysis using the density functional theory (DFT) method predicted the bowl-to-bowl inversion energy barrier of sumanene to be 16.9 kcal mol^{-1}. 2D-NMR analysis showed that the inversion energy barrier is 19.7–20.4 kcal mol^{-1} depending on the solvent. Nowadays, many sumanene derivatives have been synthesized and characterized. Their unique π-bowl structures and physicochemical properties make them intriguing molecular building blocks for the design and preparation of functional organic materials.

4.7.2 Carbon Nanorings and Carbon Nanobelts

The extensive application of carbon nanotubes, particularly SWCNTs, has motivated enormous interest in the synthesis of the shortest segments of their sidewalls.

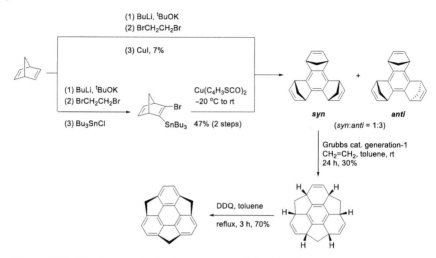

Figure 4.39 The first synthesis of sumanene by Sakurai and co-workers in 2003.

As mentioned before, CNTs produced by the arc discharge or the CVD method are mixtures with various structures and chirality. For advanced electronic and optical applications, structurally uniform SWCNTs are desired. To address this need, various covalent and non-covalent approaches for sorting out SWCNTs have been investigated. On the other hand, the bottom-up preparation of structurally uniform SWCNTs through controlled organic synthesis is a highly appealing solution to this challenge. Actually, this goal has been deemed as the Holy Grail in organic synthesis and nanocarbon chemistry. From a retrosynthetic perspective, the synthesis of a SWCNT can be achieved by using the short segment of the tube as a template or seed. This notion has considerably galvanized the synthetic efforts and studies of various **carbon nanorings** (CNRs) and **carbon nanobelts** (CNBs). Both CNRs and CNBs are macrocycles made of benzene and/or fused benzene rings connected through C–C bonds; however, they differ in the way their cyclic structures undergo ring opening. The opening of a CNR can be accomplished by breaking only one C–C bond, while a CNB requires the cleavage of at least two C–C bonds to convert it into an open form.

The shortest segment of an armchair SWCNT is a hoop of benzene rings connected through C–C bonds in their *para*-positions (Figure 4.40). This class of compounds is specifically referred to as **cycloparaphenylenes** (CPPs) [46]. The specific structure of a CPP can be denoted as [n]CPP, where n is the number of benzene rings that constitutes the nanoring. Owing to the curved π-motifs, CPPs show unique structural and electronic properties. Prior to the time when their synthesis was achieved, theoretical studies had been performed to predict that [4]CPP favors a quinoid structure on the six-membered ring, while the linkage between six-membered rings shows a

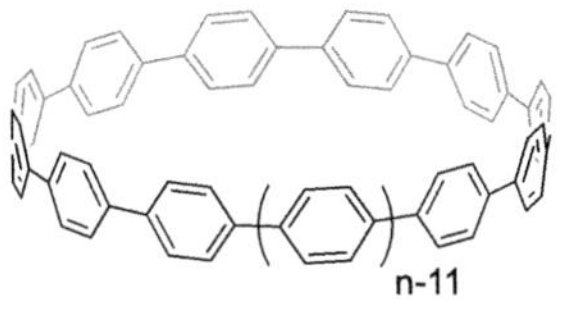

n in [n]CPP	ΔH (kcal mol^{-1})	n in [n]CPP	ΔH (kcal mol^{-1})
6	96.0	13	45.4
7	84.0	14	41.0
8	72.2	15	39.2
9	65.5	16	35.6
10	57.5	18	31.7
11	53.7	20	28.4
12	48.1		

Figure 4.40 Structure and strain energies (ΔH) of various [n]CPPs.

double bond character. This is mainly because the six-membered rings in small CPP are highly curved and hence lose a significant degree of aromaticity. As the ring size of a CPP increases, the strain energy decreases, and its π-conjugation increases. As a result, the quinoid structure gradually gives way to the benzenoid form. CPPs that contain more than five benzene rings can be considered as "true" CPPs.

The synthesis of CPPs is challenging, owing to the strain energy caused by the bending of their aromatic benzene rings. The first synthesis of CPPs was reported by Bertozzi and co-workers in 2008, who designed a strategy in which strain is built up sequentially to accomplish the assembly of the cyclic structures of CPPs [47]. In their work, a 3,6-*syn*-dimethoxy-cyclohexa-1,4-diene moiety was employed as a masked aromatic ring in a macrocyclic intermediate (Figure 4.41A). Suzuki-Miyaura cross-coupling of two curved cyclohexadiene partners afforded three macrocyclized products, which were predisposed to form [9], [12], and [18] CPPs upon efficient aromatization enabled by lithium naphthalenide. Figure 4.41B depicts the mechanism for this aromatization reaction, where the reaction begins with a one-electron transfer from lithium napthalenide to the dimethoxycyclo-hexadiene moiety, cleaving one of the C–O bonds to form a stabilized radical. A second step of electron transfer followed by elimination of the other methoxy group leads to the formation of an aromatic benzene ring.

After Bertozzi's synthesis of CPP, similar strain building up strategies have been employed by other research groups to selectively construct various CCPs. Figure 4.42A illustrates a selective synthesis of [12]CPP developed by Itami and co-work-ers in 2009 [48]. In this synthesis, they first used a Suzuki-Miyaura cross-coupling reaction to form a macrocycle under optimized conditions using the Buchwald XPhos ligand. The macrocyclization achieved a single cyclic product with a good yield of 51%. Dihydroxycyclohexane and methoxymethyl (MOM) protected dihy-droxycyclohexane units are embedded in the macrocycle as precursors for aroma-tization. In the last step of synthesis, dehydration and oxidation reactions occurred on each of the cyclohexane unit upon treatment with a strong organic acid, *p*-tol-uenesulfonic acid (*p*-TsOH), in *m*-xylene at 150 °C under microwave irradiation. This aromatization method effectively led to the formation of [12]CPP in a very good yield. In 2010, Yamago and co-workers reported a "square to loop" synthetic

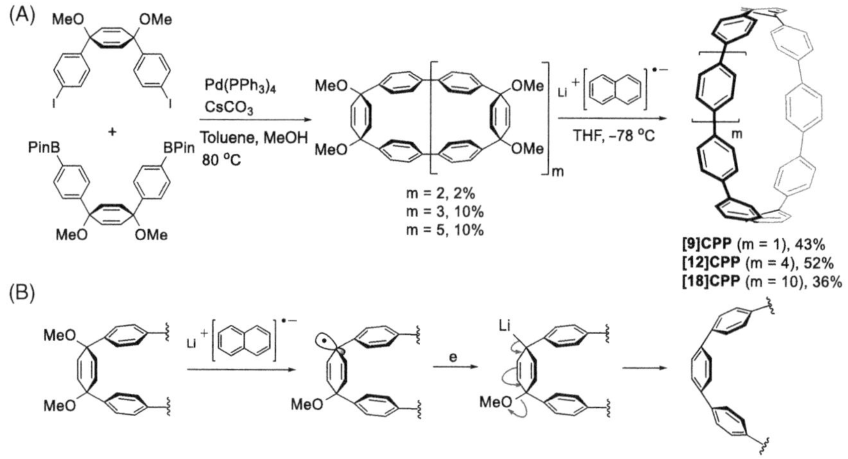

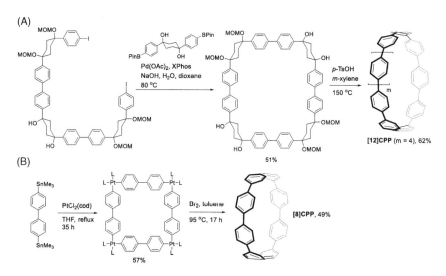

Figure 4.41 (A) The synthetic route to [9], [12], and [18]CPPs reported by Bertozzi and co-workers in 2008. (B) Proposed mechanism for the reductive aromatization of dimethoxycyclohexadiene.

Figure 4.42 (A) Selective synthesis of [12]CPP by Itami and co-workers in 2009. (B) Synthesis of [8]CPP by Yamago and co-workers in 2010.

approach for [8]CPP (Figure 4.42B) [49]. Their synthesis involved the formation of a square-shaped platinum biphenyl intermediate, which subsequently underwent bromine-induced reductive elimination at elevated temperatures to yield [8] CPP in 49% yield.

Following the pioneering works just mentioned, the chemistry of CPPs has been rapidly expanded with more complex structures synthesized. For example, the family of [n]CPPs now has expanded from $n = 7$ to 16. Naphthalene and other arene groups have been introduced to CPP structures to form structures resembling the segments of chiral CNTs. Heteroatoms such as nitrogen, oxygen, and sulfur have been incorporated into CPPs to yield new CNRs with modified electronic and chemical properties. Since these novel nanocarbons were successfully synthesized, their applications in sensing, bioimaging, electronic, and luminescence devices have been continuously explored. It is noteworthy that an elegant work in which CPPs were used to synthesize SWCNTs was presented by Itami and co-workers in 2013 (see Figure 4.43) [50]. Herein a selective growth of structurally uniform carbon nanotubes was initiated using [12]CPP as a template. Ethanol was used as the carbon source and the nanotubes were grown through the CVD method. The average diameter of the carbon nanotubes obtained was observed to be close to that of [12]CPP. This work represents a cornerstone in the bottom-up synthesis of structurally uniform CNTs.

Alkynyl units can be incorporated into molecular hoops to yield intriguing CNRs as well. Figure 4.44 illustrates two examples of cyclophenyleneethynylenes, which well represent this family of carbon nanorings. Unlike the benzene ring, the bending of an alkynyl sp hybridized carbon is much easier. Alkyne formation can be readily achieved through elimination reactions as well as metal-catalyzed coupling and metathesis reactions, which offer efficient synthetic tools for the construction of this type of CNRs. As a matter of fact, the synthesis of [6]CPPA, one of the earliest members of the CNR family, was accomplished by Kawase and Oda in 1996 [51]. It is worth noting that the alkyne-based CNRs possess a well-defined cavity that is suitable for the formation of supramolecular host–guest complexes with various fullerenes.

A nanoring made of all sp hybridized carbon atoms was synthesized by Anderson, Gross, and co-workers in 2020 [52]. As shown in Figure 4.45, a cyclic oligo(enyne)

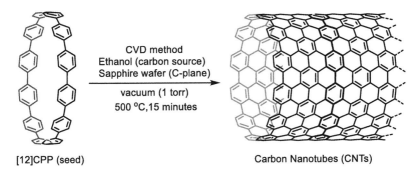

[12]CPP (seed)   Carbon Nanotubes (CNTs)

CVD method
Ethanol (carbon source)
Sapphire wafer (C-plane)
vacuum (1 torr)
500 °C, 15 minutes

Figure 4.43 Synthesis of structurally uniform SWCNTs using [12]CPP as the seed (drawings by courtesy of Prof. Kenichiro Itami).

intermediate ($C_{18}Br_6$) was first prepared through a Cu(I)-catalyzed alkyne homocoupling reaction. This intermediate carries six bromo groups on its three alkenyl units, which can be readily converted into alkynyl units through dehalogenation on a sodium chloride bilayer on a Cu(111) substrate at 5 K in 64% yield. The formation of the cyclo[18]carbon was confirmed by high-resolution atomic force microscopic analysis.

Compared with CNRs, CNBs are more challenging synthetic targets owing to their high strain energies and the high expected reactivity for certain types of CNBs. Theoretical studies of CNBs started as early as 1954, when Heilbronner investigated the orbital properties of a class of hypothetical [n]cyclacenes [53]. [n] Cyclacenes represent the thinnest slices of zigzag CNTs (Figure 4.46). It is also noteworthy that [n]cyclacenes show isolated benzene sextets in their structures.

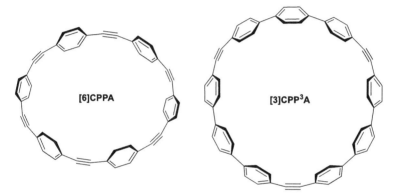

Figure 4.44 Representative structures of cyclophenyleneethynylenes.

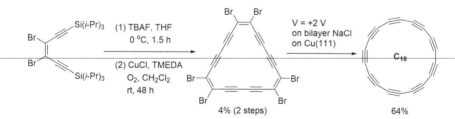

Figure 4.45 Synthesis of cyclo[18]carbon by Anderson and co-workers.

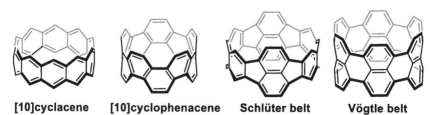

[10]cyclacene   **[10]cyclophenacene**   **Schlüter belt**   **Vögtle belt**

Figure 4.46 Structures of various aromatic belts proposed in the early years.

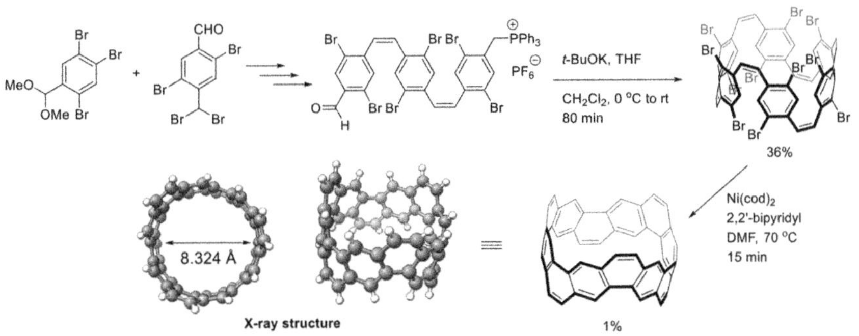

Figure 4.47 The first synthesis of a CNB by Itami and co-workers in 2017.

According to the Clar's rule, [n]cyclacenes are predicted to show poor aromaticity-stabilization and hence high reactivity. The **Clar's rule** or **Clar's π-sextet rule** was formulated by an Austrian organic chemist, Erich J. Clar, in his book *The Aromatic Sextet* published in 1972. Clar's rule states that the Kekulé resonance structure with the largest number of disjointed aromatic π-sextets (benzene-like moieties) is the most important factor governing the properties of polycyclic aromatic hydrocarbons (PAHs). Apart from [n]cyclacenes, other intriguing CNBs were also proposed, which were referred to as aromatic belts in the studies early on [54]. Among them, the [n]cyclophenacenes that represent the shortest segments of armchair carbon nanotubes, the Schlüter belt that resembles the equator region of fullerenes, and the Vögtle belt that is a short armchair carbon nanotube, have attracted considerable synthetic attention.

Numerous attempts to synthesize CNBs have been actively made for several decades after the initial theoretical proposal and investigations of CNBs. However, most failed to produce CNBs until 2017 when Itami and co-workers reported the first synthesis of a CNB (Figure 4.47) [55]. In their synthesis, an all-*cis*-benzannulene precursor was prepared initially using Wittig olefination reactions. Each benzene ring in this precursor carries two bromine groups, which are predisposed to undergo intramolecular ring cyclization reactions through the Ni-mediated Yamamoto coupling reactions. By using this elegantly designed strategy, Itami and co-workers successfully synthesized and isolated a CNB that represents the segment of (6,6) CNT in a yield of 1%. Remarkably, the molecular structure of this CNB was characterized by X-ray structural analysis, and the CNB was observed to show red fluorescence in solution and the solid state. Itami's ground-breaking synthesis soon opened the floodgates of CNB synthetic chemistry. Just within a few years after Itami's works, a variety of CNBs including chiral CNBs have been prepared and reported.

4.7.3 Molecular Nanographenes and Graphene Nanoribbons

Pristine graphene has a planar structure with a zero energy band gap, which is responsible for its semi-metallic properties. Nonetheless, the lack of a band gap hinders the application of graphene in certain electronic applications such as field-effect transistors (FETs). Nanographenes are structural segments of graphene with dimensions on the nanometer scale. They show nonzero band gaps and tunable electronic properties, depending on their specific structures. Moreover, by introducing non-six-membered rings or helical units, nanographenes can be imparted with chirality. They show chiroptical properties and hence find more extensive optoelectronic applications.

The bottom-up chemical synthesis of graphene provides access to various structurally defined molecular nanographenes with monodispersed, defect-free molecular structures. Polycyclic aromatic hydrocarbons (PAHs) that consist of sp^2 carbon frameworks with dimensions larger than 1 nm can be viewed as the smallest possible nanographenes. Pioneering work on the synthesis and characterization of such molecules can be dated back as early as 1910, when Roland Scholl, a Swiss chemist, found that aromatic compounds can undergo C–C bond forming reactions in the presence of strong Lewis acids or oxidants, such as the examples illustrated in Figure 4.48. These types of reactions are known as the **Scholl reactions**, which are nowadays the workhorse reactions for the synthesis of various nanographenes and related structures [56].

The Scholl reactions can be defined as the elimination of two aryl-bound hydrogens accompanied by the formation of an aryl–aryl bond under the influence of Friedel-Crafts catalysts. Mechanistic studies have disclosed that arenenium

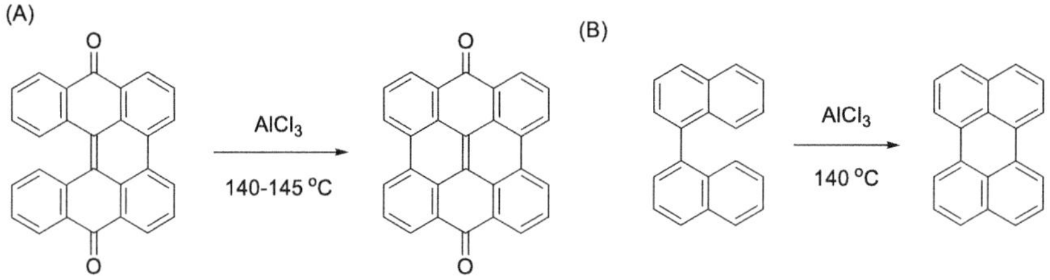

Figure 4.48 Examples of the Scholl reactions.

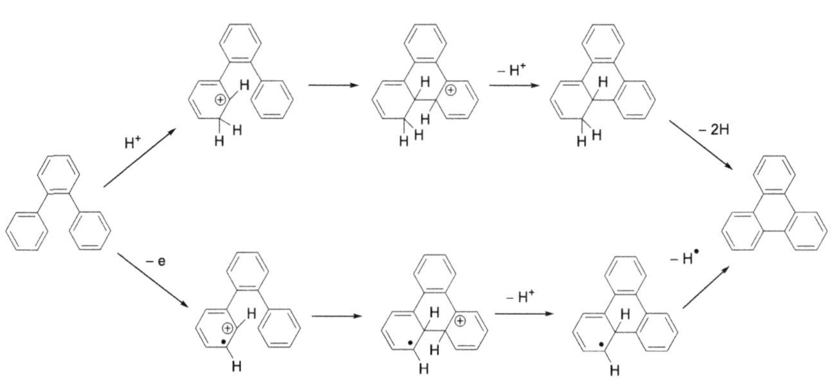

Figure 4.49 Two plausible mechanistic pathways for a Scholl reaction.

cations and radical cations (see Figure 4.49) are the key reactive species in the reaction mechanisms [57].

The Scholl reactions have been extensively applied in the synthesis of various molecular nanographenes. Figure 4.50 illustrates a remarkable synthesis of a large planar PAH reported by Müllen and co-worker in 2002 [58]. In their work, an oligophenylene precursor was prepared and then subjected to the Scholl reaction to form a molecular nanographene ($C_{222}H_{42}$). A total of 180 hydrogen atoms were eliminated in this step, testifying to the high efficiency of the reaction.

One-dimensional extension of molecular nanographene can lead to ribbon-shaped nanographenes with high aspect ratios, which are called **graphene nanoribbons** (GNRs). Figure 4.51 illustrates the synthesis of GNRs reported by Müllen and co-workers in 2014, who applied the Scholl reaction to a phenylene polymer to form well-defined, liquid-phase processable GNRs [59].

In addition to the Scholl reactions, the synthesis of nanographenes and GNRs have also been achieved in recent years by using other types of reactions, such as metal-mediated coupling [60], cyclodehalogenation [61], and benzannulation reactions (see Figure 4.52) [62].

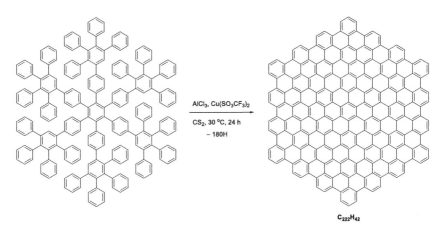

AlCl$_3$, Cu(SO$_3$CF$_3$)$_2$

CS$_2$, 30 °C, 24 h

− 180H

C$_{222}$H$_{42}$

Figure 4.50 Synthesis of a planar molecular nanographene through the Scholl reaction.

FeCl$_3$, CH$_2$Cl$_2$
MeNO$_2$, 3 days

89%

R = C$_{12}$H$_{25}$

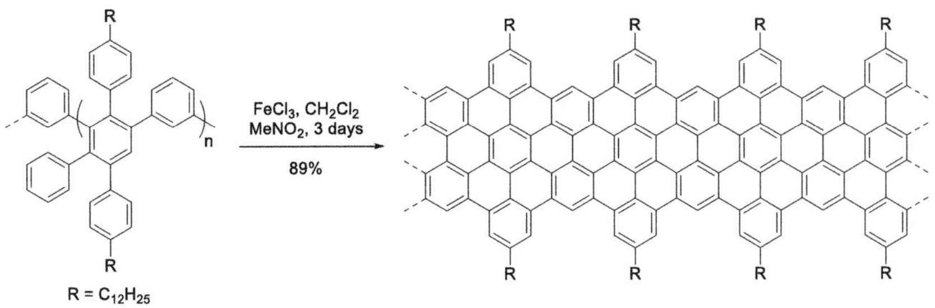

Figure 4.51 Synthesis of GNRs through the Scholl reaction.

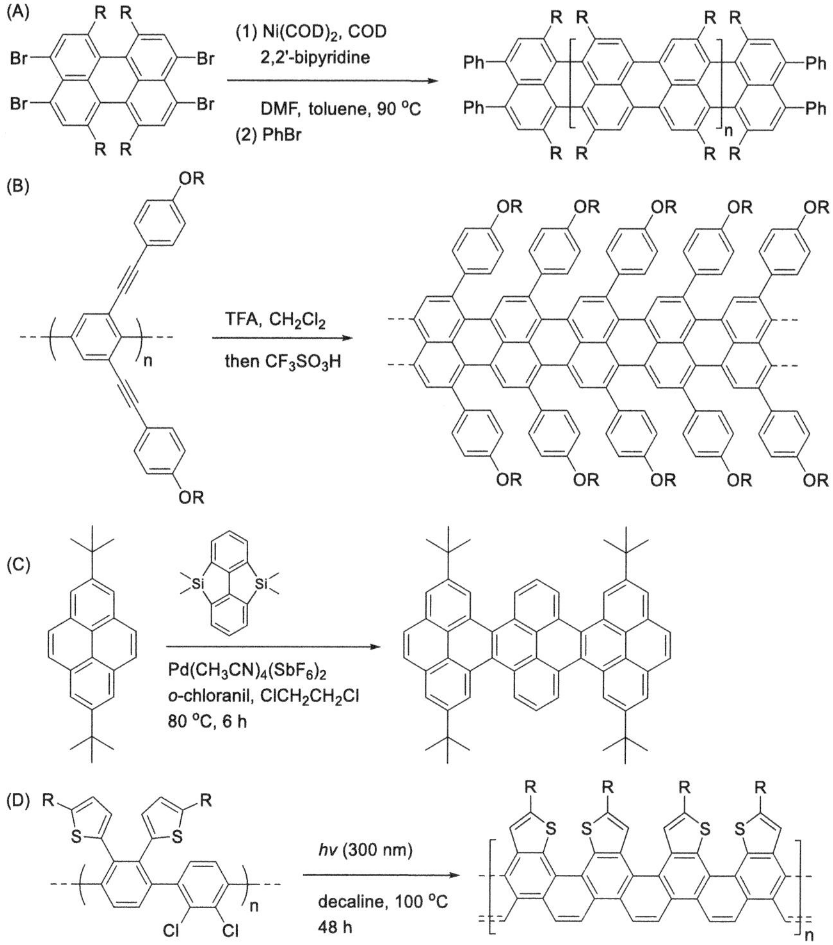

Figure 4.52 Synthetic methods recently developed for the preparation of nanographenes and GNRs.

Further Reading

- Akasaka, T.; Wudl, F.; Nagase, S., Chemistry of Nanocarbons. John Wiley & Sons: West Sussex, 2010.
- Marcaccio, M.; Paolucci, F., Making and Exploiting Fullerenes, Graphene, and Carbon Nanotubes. Springer: Berlin, 2014.
- De La Puente, F. L.; Nierengarten, J.-F., Fullerenes: Principles and Applications. 2nd ed.; Royal Society of Chemistry: Cambridge, UK, 2011.

- Li, Y.; Maruyama, S., Single-Walled Carbon Nanotubes: Preparation, Properties and Applications. Springer: 2019.
- Jiang, D.-e.; Chen, Z., Graphene Chemistry: Theoretical Perspectives. John Wiley & Sons, Ltd.: West Sussex, UK, 2013.

References

1 (a) Kroto, H. W.; Heath, J. R.; O'Brien, S. C.; Curl, R. F.; Smalley, R. E., C60: Buckminsterfullerene. *Nature* **1985**, *318*, 162–163; (b) Smalley, R. E., Discovering the Fullerenes (Nobel Lecture). *Angew. Chem. Int. Ed. Engl.* **1997**, *36*, 1594-1601.

2 Prinzbach, H.; Weiler, A.; Landenberger, P.; Wahl, F.; Wörth, J.; Scott, L. T.; Gelmont, M.; Olevano, D.; V. Issendorff, B. Gas-Phase Production and Photoelectron Spectroscopy of the Smallest Fullerene, C20. *Nature* **2000**, *407*, 60–63.

3 Krätschmer, W.; Lamb, L. D.; Fostiropoulos, K.; Huffman, D. R. Solid C60: A New Form of Carbon. *Nature* **1990**, *347*, 354–358.

4 Iijima, S.; Yudasaka, M.; Yamada, R.; Bandow, S.; Suenaga, K.; Kokai, F.; Takahashi, K. Nano-Aggregates of Single-Walled Graphitic Carbon Nano-Horns. *Chem. Phys. Lett.* **1999**, *309*, 165–170.

5 Ugarte, D. Curling and Closure of Graphitic Networks under Electron-Beam Irradiation. *Nature* **1992**, *359*, 707–709.

6 Georgakilas, V.; Perman, J. A.; Tucek, J.; Zboril, R. Broad Family of Carbon Nanoallotropes: Classification, Chemistry, and Applications of Fullerenes, Carbon Dots, Nanotubes, Graphene, Nanodiamonds, and Combined Superstructures. *Chem. Rev.* **2015**, *115*, 4744–4822.

7 Wallace, P. R. The Band Theory of Graphite. *Phys. Rev.* **1947**, *71*, 622.

8 Boehm, H.; Setton, R.; Stumpp, E. Nomenclature and Terminology of Graphite Intercalation Compounds. *Carbon* **1986**, *24*, 241–245.

9 Meyer, J. C.; Geim, A. K.; Katsnelson, M. I.; Novoselov, K. S.; Booth, T. J.; Roth, S. The Structure of Suspended Graphene Sheets. *Nature* **2007**, *446*, 60–63.

10 Narita, A.; Wang, X.-Y.; Feng, X.; Müllen, K. New Advances in Nanographene Chemistry. *Chem. Soc. Rev.* **2015**, *44*, 6616–6643.

11 Osawa, E. Superaromaticity. *Kagaku* **1970**, *25*, 854–863.

12 Murayama, H.; Tomonoh, S.; Alford, J. M.; Karpuk, M. E. Fullerene Production in Tons and More: From Science to Industry. *Fuller. Nanotub. Carbon Nanostructures* **2005**, *12*, 1–9.

13 Sawamura, M.; Iikura, H.; Nakamura, E. The First Pentahaptofullerene Metal Complexes. *J. Am. Chem. Soc.* **1996**, *118*, 12850–12851.

14 Bingel, C. Cyclopropanierung von Fullerenen. *Chem. Ber.* **1993**, *126*, 1957–1959.

15 Thilgen, C.; Diederich, F. Tether-Directed Remote Functionalization of Fullerenes C60 and C70. *Comptes Rendus Chim.* **2006**, *9*, 868–880.

16 Biglova, Y. N. [2 + 1] Cycloaddition Reactions of Fullerene C60 Based on Diazo Compounds. *Beilstein J. Org. Chem.* **2021**, *17*, 630–670.

17 Maggini, M.; Scorrano, G.; Prato, M. Addition of Azomethine Ylides to C60: Synthesis, Characterization, and Functionalization of Fullerene Pyrrolidines. *J. Am. Chem. Soc.* **1993**, *115*, 9798–9799.

18 Georgakilas, V.; Kordatos, K.; Prato, M.; Guldi, D. M.; Holzinger, M.; Hirsch, A. Organic Functionalization of Carbon Nanotubes. *J. Am. Chem. Soc.* **2002**, *124*, 760–761.

19 Śliwa, W. Diels-Alder Reactions of Fullerenes. *Fullerene Sci. Technol.* **1997**, *5*, 1133–1175.

20 Komatsu, K.; Murata, M.; Murata, Y. Encapsulation of Molecular Hydrogen in Fullerene C_{60} by Organic Synthesis. *Science* **2005**, *307*, 238–240.

21 Scott, L. T.; Hashemi, M. M.; Meyer, D. T.; Warren, H. B. Corannulene. A Convenient New Synthesis. *J. Am. Chem. Soc.* **1991**, *113*, 7082–7084.

22 Scott, L. T.; Boorum, M. M.; McMahon, B. J.; Hagen, S.; Mack, J.; Blank, J.; Wegner, H.; de Meijere, A. A Rational Chemical Synthesis of C60. *Science* **2002**, *295*, 1500–1503.

23 Boorum, M. M.; Vasil'ev, Y. V.; Drewello, T.; Scott, L. T. Groundwork for a Rational Synthesis of C60: Cyclodehydrogenation of a C60H30 Polyarene. *Science* **2001**, *294*, 828–831.

24 Ghosh, S.; Bachilo, S. M.; Weisman, R. B. Advanced Sorting of Single-Walled Carbon Nanotubes by Nonlinear Density-Gradient Ultracentrifugation. *Nat. Nanotechnol.* **2010**, *5*, 443–450.

25 Tu, X.; Manohar, S.; Jagota, A.; Zheng, M. DNA Sequence Motifs for Structure-Specific Recognition and Separation of Carbon Nanotubes. *Nature* **2009**, *460*, 250–253.

26 Johnson, R. R.; Kohlmeyer, A.; Johnson, A. C.; Klein, M. L. Free Energy Landscape of a DNA−Carbon Nanotube Hybrid Using Replica Exchange Molecular Dynamics. *Nano Lett.* **2009**, *9*, 537–541.

27 Kang, Y. K.; Lee, O.-S.; Deria, P.; Kim, S. H.; Park, T.-H.; Bonnell, D. A.; Saven, J. G.; Therien, M. J. Helical Wrapping of Single-Walled Carbon Nanotubes by Water Soluble Poly (*P*-phenyleneethynylene). *Nano Lett.* **2009**, *9*, 1414–1418.

28 Liang, S.; Zhao, Y.; Adronov, A. Selective and Reversible Noncovalent Functionalization of Single-Walled Carbon Nanotubes by a pH-Responsive Vinylogous Tetrathiafulvalene–Fluorene Copolymer. *J. Am. Chem. Soc.* **2014**, *136*, 970–977.

29 Chen, R. J.; Zhang, Y.; Wang, D.; Dai, H. Noncovalent Sidewall Functionalization of Single-Walled Carbon Nanotubes for Protein Ommobilization. *J. Am. Chem. Soc.* **2001**, *123*, 3838–3839.

30 Liang, F.; Sadana, A. K.; Peera, A.; Chattopadhyay, J.; Gu, Z.; Hauge, R. H.; Billups, W. E. A Convenient Route to Functionalized Carbon Nanotubes. *Nano Lett.* **2004**, *4*, 1257–1260.

31 Mickelson, E. T.; Huffman, C. B.; Rinzler, A. G.; Smalley, R. E.; Hauge, R. H.; Margrave, J. L. Fluorination of Single-wall Carbon Nanotubes. *Chem. Phys. Lett.* **1998**, *296*, 188–194.

32 (a) Boul, P.; Liu, J.; Mickelson, E.; Huffman, C.; Ericson, L.; Chiang, I.; Smith, K.; Colbert, D.; Hauge, R.; Margrave, J., Reversible Sidewall Functionalization of Buckytubes. *Chem. Phys. Lett.* **1999**, *310*, 367–372; (b) Mickelson, E.; Chiang, I.; Zimmerman, J.; Boul, P.; Lozano, J.; Liu, J.; Smalley, R.; Hauge, R.; Margrave, J., Solvation of Fluorinated Single-Wall Carbon Nanotubes in Alcohol Solvents. *J. Phys. Chem. B* **1999**, *103*, 4318-4322.

33 Strano, M. S.; Dyke, C. A.; Usrey, M. L.; Barone, P. W.; Allen, M. J.; Shan, H.; Kittrell, C.; Hauge, R. H.; Tour, J. M.; Smalley, R. E. Electronic Structure Control of Single-Walled Carbon Nanotube Functionalization. *Science* **2003**, *301*, 1519–1522.

34 Hirsch, A.; Englert, J. M.; Hauke, F. Wet Chemical Functionalization of Graphene. *Acc. Chem. Res.* **2013**, *46*, 87–96.

35 Hummers, W. S.; Offeman, R. E. Preparation of Graphitic Oxide. *J. Am. Chem. Soc.* **1958**, *80*, 1339-1339.

36 Li, W.; Li, Y.; Xu, K. Azidated Graphene: Direct Azidation from Monolayers, Click Chemistry, and Bulk Production from Graphite. *Nano Lett.* **2019**, *20*, 534–539.

37 Qu, S.; Li, M.; Xie, L.; Huang, X.; Yang, J.; Wang, N.; Yang, S. Noncovalent Functionalization of Graphene Attaching [6,6]-phenyl-c61-butyric Acid Methyl Ester (PCBM) and Application as Electron Extraction Layer of Polymer Solar Cells. *ACS Nano* **2013**, *7*, 4070–4081.

38 Xu, X.; Ray, R.; Gu, Y.; Ploehn, H. J.; Gearheart, L.; Raker, K.; Scrivens, W. A. Electrophoretic Analysis and Purification of Fluorescent Single-Walled Carbon Nanotube Fragments. *J. Am. Chem. Soc.* **2004**, *126*, 12736–12737.

39 (a) Lim, S. Y.; Shen, W.; Gao, Z., Carbon Quantum Dots and Their Applications. *Chem. Soc. Rev.* **2015**, *44*, 362–381; (b) Li, M.; Chen, T.; Gooding, J. J.; Liu, J., Review of Carbon and Graphene Quantum Dots for Sensing. *ACS Sensors* **2019**, *4*, 1732-1748; (c) Kumar, P.; Dua, S.; Kaur, R.; Kumar, M.; Bhatt, G., A Review on Advancements in Carbon Quantum Dots and Their Application in Photovoltaics. *RSC Adv.* **2022**, *12*, 4714-4759.

40 Sun, C. H.; Lu, G. Q.; Cheng, H. M. Nonplanar Distortions and Strain Energies of Polycyclic Aromatic Hydrocarbons. *J. Phys. Chem. B* **2006**, *110*, 4563–4568.

41 Dobrowolski, M. A.; Ciesielski, A.; Cyrański, M. K. On the Aromatic Stabilization of Corannulene and Coronene. *Phys. Chem. Chem. Phys.* **2011**, *13*, 20557–20563.

42 Lawton, R. G.; Barth, W. E. Synthesis of Corannulene. *J. Am. Chem. Soc.* **1971**, *93*, 1730–1745.

43 Butterfield, A. M.; Gilomen, B.; Siegel, J. S. Kilogram-Scale Production of Corannulene. *Org. Proc. Res. Dev.* **2012**, *16*, 664–676.

44 Mehta, G.; Shahk, S. R.; Ravikumarc, K. Towards the Design of Tricyclopenta[def, Jkl, Pqr] Triphenylene ('sumanene'): A 'Bowl-shaped' Hydrocarbon Featuring A Structural Motif Present in C60 (Buckminsterfullerene). *J. Chem. Soc., Chem. Commun.* **1993**, 1006–1008.

45 Sakurai, H.; Daiko, T.; Hirao, T. A Synthesis of Sumanene, A Fullerene Fragment. *Science* **2003**, *301*, 1878-1878.

46 Omachi, H.; Segawa, Y.; Itami, K. Synthesis of Cycloparaphenylenes and Related Carbon Nanorings: A Step toward the Controlled Synthesis of Carbon Nanotubes. *Acc. Chem. Res.* **2012**, *45*, 1378–1389.

47 Jasti, R.; Bhattacharjee, J.; Neaton, J. B.; Bertozzi, C. R. Synthesis, Characterization, and Theory of [9]-,[12]-, and [18] Cycloparaphenylene: Carbon Nanohoop Structures. *J. Am. Chem. Soc.* **2008**, *130*, 17646–17647.

48 Takaba, H.; Omachi, H.; Yamamoto, Y.; Bouffard, J.; Itami, K. Selective Synthesis of [12]cycloparaphenylene. *Angew. Chem. Int. Ed.* **2009**, *48*, 6112–6116.

49 Yamago, S.; Watanabe, Y.; Iwamoto, T. Synthesis of [8]cycloparaphenylene from a Square-Shaped Tetranuclear Platinum Complex. *Angew. Chem. Int. Ed.* **2010**, *49*, 757–759.

50 Omachi, H.; Nakayama, T.; Takahashi, E.; Segawa, Y.; Itami, K. Initiation of Carbon Nanotube Growth by Well-Defined Carbon Nanorings. *Nat. Chem.* **2013**, *5*, 572–576.

51 Kawase, T.; Darabi, H. R.; Oda, M. Cyclic [6]-and [8]paraphenylacetylenes. *Angew. Chem. Int. Ed. Engl.* **1996**, *35*, 2664–2666.

52 Scriven, L. M.; Kaiser, K.; Schulz, F.; Sterling, A. J.; Woltering, S. L.; Gawel, P.; Christensen, K. E.; Anderson, H. L.; Gross, L. Synthesis of Cyclo[18] carbon via Debromination of C18Br6. *J. Am. Chem. Soc.* **2020**, *142*, 12921–12924.

53 Heilbronner, E. Molecular Orbitals in homologen Reihen Mehrkerniger Aromatischer Kohlenwasserstoffe: I. Die Eigenwerte yon LCAO-MO's in Homologen Reihen. *Helv. Chim. Acta* **1954**, *37*, 921–935.

54 Eisenberg, D.; Shenhar, R.; Rabinovitz, M. Synthetic Approaches to Aromatic Belts: Building up Strain in Macrocyclic Polyarenes. *Chem. Soc. Rev.* **2010**, *39*, 2879–2890.

55 Povie, G.; Segawa, Y.; Nishihara, T.; Miyauchi, Y.; Itami, K. Synthesis of a Carbon Nanobelt. *Science* **2017**, *356*, 172–175.

56 Jassas, R. S.; Mughal, E. U.; Sadiq, A.; Alsantali, R. I.; Al-Rooqi, M. M.; Naeem, N.; Moussa, Z.; Ahmed, S. A. Scholl Reaction as a Powerful Tool for the Synthesis of Nanographenes: A Systematic Review. *RSC Adv.* **2021**, *11*, 32158–32202.

57 Grzybowski, M.; Skonieczny, K.; Butenschön, H.; Gryko, D. T. Comparison of Oxidative Aromatic Coupling and the Scholl Reaction. *Angew. Chem. Int. Ed.* **2013**, *52*, 9900–9930.

58 Simpson, C. D.; Brand, J. D.; Berresheim, A. J.; Przybilla, L.; Räder, H. J.; Müllen, K. Synthesis of a Giant 222 Carbon Graphite Sheet. *Chem. Eur. J.* **2002**, *8*, 1424–1429.

59 Narita, A.; Feng, X.; Hernandez, Y.; Jensen, S. A.; Bonn, M.; Yang, H.; Verzhbitskiy, I. A.; Casiraghi, C.; Hansen, M. R.; Koch, A. H. R.; Fytas, G.; Ivasenko, O.; Li, B.; Mali, K. S.; Balandina, T.; Mahesh, S.; De Feyter, S.; Müllen, K. Synthesis of Structurally Well-Defined and Liquid-Phase-Processable Graphene Nanoribbons. *Nat. Chem.* **2014**, *6*, 126–132.

60 Ozaki, K.; Kawasumi, K.; Shibata, M.; Ito, H.; Itami, K. One-Shot K-Region-Selective Annulative π-Extension for Nanographene Synthesis and Functionalization. *Nat. Commun.* **2015**, *6*, 6251.

61 Jolly, A.; Miao, D.; Daigle, M.; Morin, J.-F. Emerging Bottom-Up Strategies for the Synthesis of Graphene Nanoribbons and Related Structures. *Angew. Chem. Int. Ed.* **2020**, *59*, 4624–4633.

62 Senese, A. D.; Chalifoux, W. A. Nanographene and Graphene Nanoribbon Synthesis via Alkyne Benzannulations. *Molecules* **2018**, *24*, 118.

5

Synthetic Molecular Devices and Machines

5.1 Basic Concepts of Molecular Devices and Machines

As already mentioned in Chapter 1, the visionary prediction of ultraminiaturized devices and machines by Richard Feynman in his famous lecture, *There is plenty room at the bottom*, marked the beginning of modern nanotechnology and nanoscience. More than a half century has passed ever since Feynman's prophecy made in 1959. Now the dream of molecular machines has not only been proven feasible but fruitfully realized through the concerted efforts made by scientific researchers across multiple disciplines. Among them, synthetic organic chemists have made critically important contributions to the design and assembly of diverse types of molecular devices and machines by means of innovative organic syntheses and characterizations.

So, what are "molecular devices" and "molecular machines"? How do they differ from other types of functional molecules and supramolecular systems? To answer these questions, we need to first clarify the concepts of devices and machines. According to the Encyclopedia Britannica, a machine or a device has "a unique purpose, that augments or replaces human or animal effort for the accomplishment of physical tasks." Our macroscopic world is full of various types of devices and machines, which utilize, apply, and transmit different forms of energies (e.g., fossil fuel, electricity, sunlight, wind, hydraulic and nuclear power). Depending on their purposes, devices and machines can be big or small, simple or complex. Generally speaking, a device or a machine contains different components, each of which is designed for a simple specific function. The combination of these components allows more complex and useful functions to be achieved. For instance, the structure of an electric motor consists of two basic parts – a rotor and a stator. The rotor is made up of coiled wires and is situated in the middle of the stator that is lined with magnets or with coil windings. The motor operates through the interaction between

Organic Nanochemistry: From Fundamental Concepts to Experimental Practice,
First Edition. Yuming Zhao.
© 2024 John Wiley & Sons, Inc. Published 2024 by John Wiley & Sons, Inc.

the motor's magnetic field and an electric current to convert electrical energy into mechanical energy. In contrast, a Boeing 777 wide-body airliner is far more complex machine. It contains approximately 3 million parts and takes a Boeing team about 46 days to assemble a single airplane. Operating with aviation fuel, this engineering marvel can carry more than 300 passengers and fly a distance up to 9700 km.

The concepts of devices and machines can be applied to the molecular level. A molecular machine can be described as an assembly of discrete molecular components, in which each component performs a simple task. The entire molecular system delivers more complex function through the cooperation of the component parts. While macroscopic devices and machines are made of various types of materials (e.g., metals, plastics, wood, textiles), molecular devices and machines can only be made from molecules through the formation of various covalent and non-covalent bonds. So, the construction of a molecular machine in practice has no difference than the synthesis of a pharmaceutical compound or a natural product, except for the structure and purpose of the target molecule. At the molecular level, the operation of a machine needs to be driven by energies that can effectively interact with a molecular or a supramolecular system to induce desired electronic and nuclear rearrangement. These include thermal activation, chemical reactions, electron transfer, and photon excitation. It is also worth mentioning that, in contrast to bulk materials, the properties of molecular machines on the nanoscale are governed by the laws of quantum mechanics. In the nanoworld, electromagnetic interactions and interference and confinement of electron waves (i.e., quantum effects) are critical, making molecular entities somewhat floppy and sticky when they are engaged with one another. Moreover, gravity and inertia become unimportant at the nanoscale. As a result of these properties, normal engineering intuition fails to provide meaningful guidance to the design of devices and machines which are extremely small. Newtonian mechanics gradually give way to quantum mechanics as machines become increasingly miniaturized. This concept is well known as the **scaling laws**. Simply put, a molecular entity that resembles the shape and construct of a macroscopic mechanical device (e.g., a cogwheel or a rotor) does not necessarily yield a similar function. Considerations of molecular physics and chemistry are vitally important for the development of molecular devices and machines at the nanoscale level.

Nature has demonstrated abundant examples of nanomachines that take specific actions in all kinds of biological systems. For example, a human body contains about 10,000 different machines at work on the nanoscale to facilitate many life processes, ranging from eating, digesting, breathing to growing, reproducing, repairing, and so forth. All these biological nanomachines work in concert, making the human body exhibit the most complex mechanisms in the universe. It is also interesting to note that many of the bio-nanomachines can still retain their atom-sized functions after they are isolated and purified. Understanding of their working principles has provided great inspiration and guidance to human designed molecular machines.

Unlike macroscopic machines, nanomachines cannot be built in a smooth range of sizes from small to large. Instead, they must contain an integral number of atoms. Most of the bio-nanomachines employ proteins as building blocks and they operate in a chaotic environment, subjected to continual bombardment by water molecules. Therefore, nanomachines function very differently from those working on the macroscopic scale. For example, it is impossible to design a nanoscale molecular rotor that features a smooth ring surrounding an axle in order to undergo a smooth rotary motion. Instead, nanoscale rotary motors existing in cells work in a very different manner. Taking adenosine triphosphate (ATP) synthase as an example, it works as a molecular motor by adopting several discrete rotary states that cycle one after the other [1]. As illustrated in Figure 5.1, the structure of an ATP synthase is composed of two rotary motors, called F_0 and F_1. Component F_0 is an electric motor that is embedded in a membrane (shown schematically as a lipid bilayer). The rotation of F_0 is powered by the flow of hydrogen ions (H^+) across the membrane. The F_0 motor is connected to the second motor, termed F_1, which is a chemical motor powered by adenosine diphosphate (ADP) and inorganic phosphate. The two motors are linked together by a stator (b_2), so that when the rotation of the F_0 motor drives the turning of the second motor F_1. The rotation of F_1 then results in the formation of ATP. ATP synthase plays an

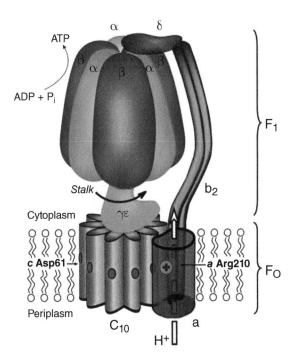

Figure 5.1 Schematic illustration of an ATP synthase embedded in a membrane. (Reproduced from *Biophys. J.* 2004, *86*, 1332–1344).

indispensable role in cellular functions, synthesizing most of the ATP that powers our cellular processes. The lessons we can learn from Nature are extremely valuable and inspirational. In the past decades, knowledge about the working principles of these natural bio-nanomachines has greatly motivated the design and synthesis of artificial molecular devices and machines to mimic the natural ones that have already been working in biological systems [2].

Prior to further discussions of various artificial molecular machines, the definitions of some important terms need to be provided. First, a **molecular machine** is defined as a molecular system that can achieve controlled motion of one molecular or sub-molecular component relative to another when triggered by an external stimulus. The operation of a molecular machine can potentially result in the accomplishment of a desired task. Molecular machines are further categorized into several classes, such as molecular switches, molecular motors, and molecular rachets. A **molecular switch** refers to a type of molecular machine in which the change in relative positions of the components influences a system as a function of the state of the switch. A **molecular motor**, however, is a molecular machine in which the change in relative position of the components influences a system as a function of the trajectory of the components. When a switch is returned to its original position, any mechanical work performed by the original switching action will be reversed (zero net work). When a molecular motor returns to its original state at the end of the rotatory cycle, net work is performed. The distinction between a molecular switch and a molecular motor is that a molecular switch cannot be used to perform work repetitively and progressively in the way that a motor does. Only molecular motors can be used to progressively drive systems away from equilibrium.

5.2 Fundamentals of Photophysical and Photochemical Principles

Energy levels of molecules are quantized and comprised of various electronic, vibrational, rotational, and translational states. When electrons in a molecule are populated in the lowest energy molecular orbitals as dictated by the Aufbau principle, the molecule is said to be in its **ground electronic state**. A ground-state molecule may be excited to a higher energy state by either absorbing a photon of light whose energy exactly matches the difference between the initial and the final state or accepting energy transferred from another molecule through collision. The absorption of a photon by a molecule may subsequently drive it to undergo a chemical reaction, and such processes are known as **photochemistry**. There are two fundamental principles which form the basis of our understanding of photochemical transformations. They are often called the two laws of photochemistry. The first one is the **Grotthuss-Draper law**, which states that light must be absorbed by a compound in order for a photochemical reaction to occur.

The second law of photochemistry is the **Stark-Einstein law**. According to this law, for each photon of light absorbed by a chemical system, only one molecule is activated for subsequent reaction. Because light is quantized, it has the properties of both a particle and a wave. An Avogadro's number of photons (i.e., one mole of light quanta) is termed an **Einstein**. The relationship between the energy of light (E) and its frequency (v) and wavelength (λ) is described by the following equation,

$$E\left(\text{kcal mol}^{-1}\right) = hv = hc / \lambda = 2.86 \times 10^4 / \lambda\left(\text{nm}\right) \tag{5.1}$$

where h is the Planck's constant ($6.62607015 \times 10^{-34}$ m^2 kg s^{-1}) and c is speed of light (approximately 3×10^{-8} m s^{-1}). According to this equation, the energy of light is linearly proportional to its frequency and inversely proportional to its wavelength. It is also convenient to express a photonic energy by wave number or reciprocal centimeters (cm^{-1}), since energy increases with increasing wave number.

Typically, photochemistry involves **photophysical processes** at the initial stage(s), in which the chemical structure of the molecule remains unchanged but the electronic and vibrational states vary. A photoirradiation in the ultraviolet-visible (UV-Vis) region of the spectrum (i.e., from 200 to 700 nm) may promote an electron in a molecule from a lower-energy orbital to a higher-energy orbital, leading to the formation of an **excited electronic state**. Such a photoexcited molecule can undergo various physical and chemical processes. A useful way to describe the energy changes in photophysical processes is called a **Jabłoński diagram**, which is named after a Polish physicist, Aleksander Jabłoński, who proposed this type of diagram in 1935.

As shown in Figure 5.2, a molecule in its ground electronic state can be promoted to an excited state through absorption of a photon. In most cases, a molecule in its ground electronic state has a closed shell electronic configuration, referred to as the S$_0$ state where S denotes a singlet spin multiplicity. The photon of light absorbed corresponds exactly to the energy difference between the ground state and the excited state. Absorption of two or more photons to give a single electronic transition generally does not happen. In principle, the absorption of a photon by a molecule leads to a so-called **vertical electronic transition**. This is well known as the **Franck–Condon principle**, which is the approximation that an electronic transition is most likely to occur without changes in the positions of the nuclei in the molecular entity and its environment. The resulting state of a vertical transition is called a **Franck–Condon state**, which has the same molecular geometry and spin multiplicity as the ground electronic state. The intensity of a vibronic transition is proportional to the square of the overlap integral between the vibrational wavefunctions of the two states that are involved in the transition. Therefore, it is common that the absorption of a photon ends up with a vibrational energy level that is not the lowest one in the electronic excited state. For example, the absorption depicted in Figure 5.2 gives a vibrationally excited

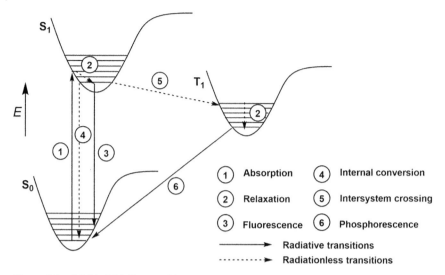

Figure 5.2 A Jabłoński diagram illustrating various photophysical processes.

state of the first electronic excited state (S_1). This state can then rapidly relax to the ground vibrational state of S_1 through a radiationless process (i.e., no photons emitted), which usually occurs by giving off heat to the surrounding medium at rates of 10^{11} to 10^{12} s^{-1}. In addition to the $S_0 \rightarrow S_1$ transition, it is also possible that photoexcitation induces transition to a higher-lying singlet state (S_n); however, most of the photophysical and photochemical processes are initiated from the lowest electronic excited states (S_1 and T_1). This has been generally known as the **Kasha's rule**. For example, the emission of a photon (termed **fluorescence**) occurs in an appreciable yield from the lowest excited state S_1. Fluorescence is a radiative transition that yields the ground electronic state (S_0), in which the vibrational level may be an excited state or the lowest energy state. Usually, the emission originates from an $S_1 \rightarrow S_0$ transition; however, **anomalous fluorescence** (e.g., $S_2 \rightarrow S_0$) does take place in some compounds. Molecules or molecular segments that give fluorescence upon photoexcitation are called **fluorophores**.

The excited state S_1 can return to the S_0 state through a radiationless process called **internal conversion** (IC). In this process, no photons are emitted. Instead, the excited state dissipates its excess energy by releasing heat to the medium. Another way to exit S_1 is to form a triplet excited state (T_n), in which two electrons spin in parallel. Usually, the S_1 state is converted into the lowest energy triplet state termed T_1. This process is called **intersystem crossing** (ISC). Note that T_1 is generally lower in energy than S_1. The transition from S_1 to T_1 may end up with an excited vibrational state in T_1. The ISC process is quantum mechanically forbidden, since the singlet and triplet wavefunctions are orthogonal. As a result, spin flipping in an organic molecule is typically slow. ISC can be facilitated by **spin–orbit coupling**, an interaction between orbital angular momentum and spin angular momentum. Spin–orbit coupling can be enhanced by the **heavy atom effect**, since heavy atoms show significant mixing of spin and orbital quantum numbers. For molecules containing only C, H, N, and O atoms, spin–orbit coupling is minimal. Nevertheless, the rate of an ISC (e.g., $S_1 \rightarrow T_1$) can be accelerated if the transition involves a change of molecular orbital type. This is known as the **El-Sayed rule**.

The conversion of an excited triplet state to the ground singlet state can be achieved through a radiative pathway called **phosphorescence**. Like fluorescence, a photon is emitted during the process of phosphorescence. However, phosphorescence involves a spin flip (e.g., $T_1 \rightarrow S_0$), so it is a quantum mechanically forbidden process. Phosphorescence is a much slower phenomenon than fluorescence. Typical fluorescence lifetimes are on the scale of nanoseconds, while phosphorescence lifetimes can be as long as seconds or even minutes. As a result, phosphorescence can continue to occur after the excitation source is turned off. Fluorescence and phosphorescence are together referred to as **luminescence**, if emission properties are discussed without any intention to distinguish between the two terms.

In a photophysical or a photochemical process, the concept of **quantum yield** (Φ) is an important indicator for determining the efficiency of the process. Quantum yield is defined by the following equation:

$$\Phi = \frac{\text{Number of molecules undergoing a process}}{\text{Number of photons absorbed}} \tag{5.2}$$

As such, the fluorescence quantum yield of a compound can be defined as the ratio between the number of photons emitted and the number of photons absorbed in fluorescence spectroscopic measurements. Quantum yield can also be correlated to relative rates. For instance, in a steady-state situation the quantum yield can be evaluated as a ratio of the rate of a photophysical or photochemical process versus the rate of photon absorption.

$$\Phi = \frac{\text{Rate of a process}}{\text{Rate of photon absorption}} \tag{5.3}$$

There are several useful guidelines for assessing the relative efficiencies of various photophysical and photochemical processes. The first factor is spin. A transition between states of different spins, such as an $S_0 \rightarrow T_1$ transition, is quantum mechanically forbidden and therefore is very inefficient. The second factor is the general spatial overlap of the wavefunctions of two states. Poor wavefunction overlap results in a lower transition efficiency, and vice versa. For example, organic chromophores usually contain π-conjugated structures, in which π-electrons can be readily photoexcited through facile transitions from π bonding to π antibonding orbitals (i.e., $\pi \rightarrow \pi^*$ transitions), since the π and π^* orbitals (wavefunctions) typically show good spatial overlap. The transition from an electron lone pair orbital to a π antibonding orbital (i.e., $n \rightarrow \pi^*$ transition), however, is inefficient due to their poor spatial overlap. The third factor that deserves serious consideration is the **energy gap law**. For transitions between two states of different spins (e.g., ISC), the energy gap law states that the smaller the energy gap between the two states, the more likely the process will take place.

When a molecule is excited through photon absorption, the resulting excited state can return to the ground state by releasing energy through radiative decay pathways (i.e., photon emission) and/or non-radiative decay pathways. Radiative decay results in fluorescence and phosphorescence. Non-radiative decay may occur through various mechanisms. When a molecule of a fluorophore in its excited state collides with another molecule, collision-induced fluorescence **quenching** may occur. It is also possible that the collision leads to the formation of a stable complex that can emit a photon. In such cases, the complexes are termed **exciplexes** or **excimers**. An excimer is a particular case of an exciplex where the two molecules are the same. For example, pyrene molecules at low concentrations show peaks at 375–405 nm in the emission spectrum. At high concentrations, there is a significant formation of excited-state pyrene dimers (excimers), exhibiting a new emission band at a wavelength around 460 nm. If a molecule contains donor (D) and acceptor (A) groups, photoinduced electron transfer and photoinduced energy transfer may occur to cause fluorescence quenching.

5.3 Photochemically Induced Olefin *Cis-Trans* Isomerization

Unlike the reactions taking place under thermal conditions, many photochemical reactions are rather complex and more challenging to understand. As discussed previously, photochemistry begins with excitation of a molecule to its electronic excited state(s); therefore, investigation of the reaction mechanism of a photochemical reaction requires the consideration of the interactions of at least two electronic states, typically S_0 and S_1. Figure 5.3 schematically depicts the process of a photoreaction, which is commonly called a **diabatic photoreaction** since there are two potential energy surfaces involved in the reaction process. Herein two reaction coordinates are shown, corresponding to the reaction pathways in the ground state (S_0) and the first excited state (S_1). The photoreaction begins with the photoexcitation of the reactant, which is an energy minimum on the potential energy surface of the S_0 state. According to the **Franck-Condon** principle, the

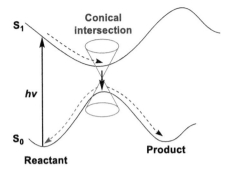

Figure 5.3 Schematic illustration of a diabatic photochemical reaction.

vertical excitation from S_0 to S_1 leads to an unstable structure on the potential energy surface of S_1. This structure then undergoes geometry changes, moving to the energy minimum of the potential energy surface of S_1. In favorable cases for photochemistry, the minimum on the excited state surface is close to the maximum on the ground state surface. The small energy gap allows the molecule to efficiently transition from S_1 to S_0 according to the energy gap law. Once it arrives at the ground state, the molecule is at an energy maximum which will spontaneously move to an energy minimum (either product or reactant) that is directly connected to it on the potential energy surface. Depending on the precise nature of the transition from one surface to another, the photoreaction can lead to the formation of a product or return to the starting material. The region where the S_0 and S_1 potential energy surfaces are close in energy is called a **funnel** or a **conical intersection** (CI), which plays a crucial role in a photoreaction as it provides the essential exiting point from the excited state to the ground state. All photochemical reactions start from the ground state and must return to the ground state.

While the diabatic mechanism is typical for most photochemical reactions, some other mechanisms may take place in certain photochemical processes. One of the possible pathways is called the **hot ground state** reaction, in which the excited state returns to the ground state through internal conversion, releasing thermal energy to drive the reaction in the ground state. Another possible pathway is the **adiabatic** reaction where the conversion of reactant to product occurs only on the surface of the excited state. The reaction is then followed by relaxation of the product from the excited state to the ground state. In this case, if the product is emissive, fluorescence of the product will be observed after excitation of the starting material.

There are numerous types of photochemical reactions that have been investigated and developed for organic synthesis. In the field of molecular devices and machines, the photoisomerization of alkenes plays a central role and hence is discussed in detail in this chapter. An alkene compound contains a C=C double bond

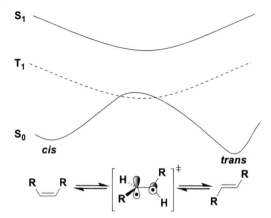

Figure 5.4 Schematic drawing of potential energy surfaces for *cis–trans* olefin isomerization.

that is restricted from free rotation in the ground state. Therefore, the *cis–trans* isomerization of an monoalkene usually needs to overcome a high activation energy barrier (around 60 kcal mol^{-1}) in the ground state. As schematically depicted in Figure 5.4, the *cis–trans* isomerization of an alkene (RCH=CHR) requires rotation about the C=C bond. When the R–C–C–R dihedral angle is rotated to 90°, the two p orbitals on the alkenyl carbons become perpendicular to one another without any interactions. As such, the twisted alkene shows a great deal of biradical character and is therefore unstable. On the ground-state potential energy surface, this geometry is the transition state, while the energy barrier to overcome it is close to the bonding energy of the π bond of an alkene.

Photoexcitation of an alkene promotes one electron in the π orbital to the antibonding π^* orbital, resulting in the formation of a π,π^* state. It is also possible that low-lying Rydberg states are present after excitation of small alkene molecules. A Rydberg state has very weakly held electrons in an orbital of exceedingly high energy. Either way, the photoexcitation of an alkene breaks a π bond. As such, the alkene molecule is free to rotate about the C=C bond as well as to pyramidalize one of the two carbon atoms in an excited state, S_1 or T_1. On the potential energy surface of S_1, the twisted structure (R–C–C–R dihedral angle = 90°) is the energy minimum because of steric and electron repulsive effects. This creates a situation where the maximum on S_0 is the minimum of S_1, leading to efficient hopping from S_1 to S_0 through a funnel or conical intersection. When the alkene molecule goes back to the ground state from a geometry close to the transition state, it follows the potential energy surface of S_0, ending up with either a *trans* or a *cis* structure as product. The last step, however, is independent of the configuration of the starting alkene in the S_0 state. The potential energy surface of T_1 is similar to that of S_1 in terms of the torsion around the C=C double bond. Compared with S_1, the surface of T_1 is much closer to S_0 in energy. In the presence of a sensitizer, the T_1 state can be more populated and the energy minimum in T_1 (twisted form) is actually lower in energy than the maximum in S_0, leading to facile *cis–trans* isomerization as well.

In the ground state, the *cis–trans* isomerization of an alkene is difficult and requires high temperatures or the use of catalysts. Under photochemical conditions, however, the *cis–trans* isomerization becomes easy and fast, due to the reasons mentioned above. As a matter of fact, both *cis* and *trans* alkenes are photoreactive. Therefore, if a pure alkene isomer (e.g., *cis*) is subjected to photoirradiation, its corresponding *trans* isomer will be generated through the *cis* → *trans* isomerization reaction. Once the *trans* isomer is formed, it can be photoexcited and reverted to the *cis* alkene isomer, assuming that the *trans* isomer also absorbs the incident light. Usually, *cis* and *trans* alkene isomers show extensively overlapped but not identical absorption spectra. In very rare cases, a particular wavelength of light can be absorbed by only one of the isomers. Therefore, typical photoirradiation will drive the alkene *cis–trans* isomerization reaction in both directions. Extensive photoirradiation can produce a **photostationary state**, in which a particular proportion of isomers does not vary upon further irradiation. At a photostationary state, the ratio between the concentrations of the two isomers under the irradiation of a single wavelength is governed by the equation shown in Figure 5.5. Herein, the relative concentration of each isomer is inversely proportional to the product of the molecular absorptivity of the isomer (ε) and the quantum yield (Φ) for its conversion to another isomer at a particular wavelength. Practically, it is possible to find a wavelength range where one isomer is more absorptive than the other, so that irradiation leading to a photostationary state will push a *cis–trans* isomerization equilibrium to favor one isomer over the other. In the example illustrated in Figure 5.5, the *trans* isomer absorbs more strongly at a relatively long wavelength λ_1. Therefore, extensive irradiation at this wavelength should make the *cis* isomer show a far greater relative population.

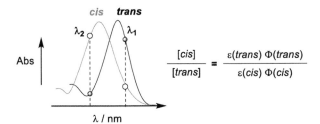

$$\frac{[cis]}{[trans]} = \frac{\varepsilon(trans)\ \Phi(trans)}{\varepsilon(cis)\ \Phi(cis)}$$

Figure 5.5 Absorption spectra of alkene *cis* and *trans* isomers used to choose appropriate wavelengths to control the ratio of isomers in a *cis–trans* photoisomerization reaction.

Irradiation at a shorter wavelength λ_2, on the other hand, will favor the formation of the *trans* isomer as the dominant one when a photostationary state is reached.

It is also worth noting that the quantum yields of the *cis* → *trans* and *trans* → *cis* conversions (i.e., $\Phi(cis)$ and $\Phi(trans)$) are mainly dependent on the potential energy surfaces of S_0 and S_1. Typically, the $\Phi(cis)$ and $\Phi(trans)$ of an alkene are close to 0.5, with the one leading to a more stable isomer being slightly greater in value.

5.4 Case Study I: Photo-driven Molecular Motors and Molecular Cars

The *cis–trans* isomerization of double bonds has been extensively studied and applied as a key working mechanism for various molecular switches and machines. Prototypical examples of such systems are stilbene and azobenzene, which have been extensively used in various photochromic molecular switches. The first example of a synthetic molecular motor capable of consecutively and unidirectionally rotating was accomplished by Feringa and co-workers in 1997 [3]. The molecular structure they synthesized, known as the first-generation molecular motor in Feringa's group, is based on an overcrowded alkene system with two stereocenters [4]. As shown in Figure 5.6, the synthesis of this molecular motor began with an α-methylation of 2,3-dihydro-1*H*-phenanthren-4-one using lithium diisopropylamide (LDA) and methyl iodide. The methylated ketone was then reduced to an alcohol with NaBH$_4$ to yield a *cis*-alcohol as the major product. The *cis*-alcohol intermediate was subjected to an esterification reaction with a chiral phthalic acid ester under the catalysis of *N,N'*-dicyclohexylcarbodiimide (DCC) and 4-dimethylaminopyridine (DMAP), resulting in two diastereomeric products

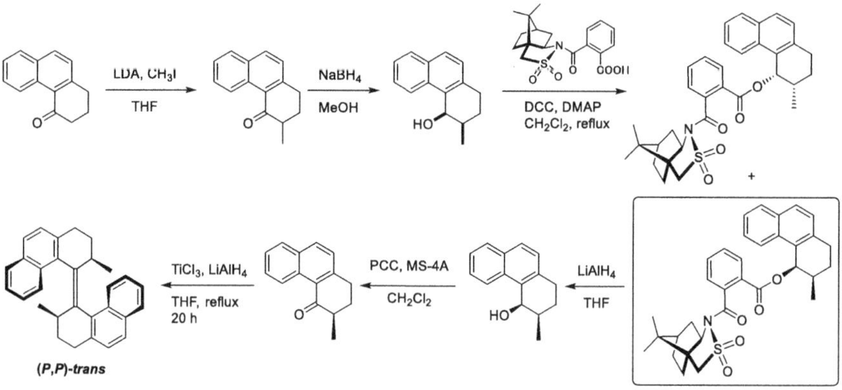

Figure 5.6 Synthesis of the first photo-driven molecular motor by Feringa and co-workers.

which could be readily separated by HPLC on silica gel. The diastereomer in which the methylated carbon has an (*R*) configuration was further reduced with LiAlH₄ and then oxidized with pyridinium chlorochromate (PCC) to afford an enantiomerically pure α-methylated ketone. Finally, the ketone was subjected to a McMurry reaction in the presence of TiCl₂ and LiAlH₄ to yield an overcrowded alkene product, named (*P,P*)-*trans* in Figure 5.6.

Structurally, the (*P,P*)-*trans* shows specific stereochemistry that enables its unique performance as a unidirectionally rotating molecular motor. The central C=C double bond takes a *trans* configuration, while the two methyl-substituted carbons adopt the (*R*) configuration. As a result of the steric interactions within this molecule, the molecule is twisted out of plane to generate a helical conformation, in which the two methyl groups take a pseudoaxial orientation and the two helical segments adopt the *P* handedness to minimize steric interactions. Upon irradiation with light λ ≥ 280 nm, the (*P,P*)-*trans* molecule undergoes a C=C double bond photoisomerization in *n*-hexane at −55 °C (step 1, in Figure 5.7). Due to the steric crowding of the two helical units, the photoisomerization favors only the clockwise rotation, yielding a *cis*-alkene, namely (*M,M*)-*cis*, as the major product at the photostationary state. Comparison of the steric bias for the two rotational directions in step 1 is illustrated in Figure 5.8A. The photoisomerization of step 1 is also reversible. Upon irradiation with λ ≥ 380 nm, the photoequilibrium

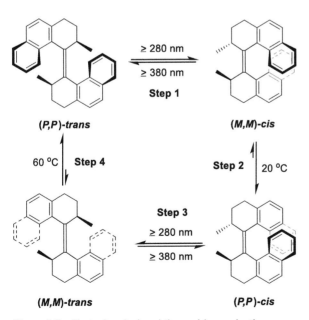

Figure 5.7 Photochemical and thermal isomerization processes of Feringa's first-generation molecular motor.

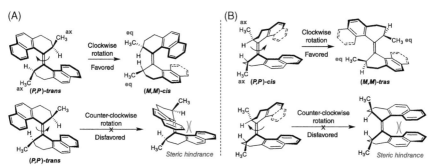

Figure 5.8 Comparison of the differences in steric hindrance for clockwise and counter-clockwise rotations of C=C bond in the photoisomerization of (A) (*P,P*)-*trans* and (B) (*P,P*)-*cis* structures.

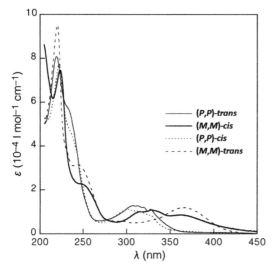

Figure 5.9 UV-Vis absorption spectra of various isomers of Feringa's first-generation molecular motor. (Reproduced from *Nature* 1999, *401*, 152–155).

shifts back to (*P,P*)-*trans* as the major product. The driving force for the reverse reaction can be found from the UV-Vis spectral differences between (*P,P*)-*trans* and (*M,M*)-*cis* isomers. As shown in Figure 5.9, the (*M,M*)-*cis* isomer absorbs more strongly in the longer wavelength region (> 350 nm) than the (*P,P*)-*trans* isomer does. According to the equation given in Figure 5.5, at wavelength greater than 380 nm, the distribution of the weakly absorbing species (*P,P*)-*trans* will be dominant when a photostationary state is reached.

It is also worth noting that the transformation from (*M,M*)-*cis* to (*P,P*)-*cis* (step 2) is called **thermal helix inversion** (THI), which is an irreversible process under the thermal conditions used experimentally due to the large energy difference between the two isomers. Even under the conditions of photoirradiation with light at λ ≥ 280 nm, the (*P,P*)-*cis* isomer will not return to the (*M,M*)-*cis* isomer. Instead, the C=C double bond of the (*P,P*)-*cis* isomer continues to undergo a *cis–trans* isomerization in a clockwise rotatory manner, leading to the formation of a (*M,M*)-*trans* isomer (step 3, Figure 5.7). The steric bias that drives the clockwise rotation in this step is shown in Figure 5.8B. Finally, (*M,M*)-*trans* can relax to the more stable structure (*P,P*)-*trans* at 60 °C. The key thermodynamic driving force for this thermal isomerization step is illustrated in Figure 5.10B. In the (*M,M*)-*trans* structure, the two methyl groups take on pseudo-equatorial positions, which incur more significant steric interactions with their neighboring groups.

The results shown in Figure 5.7 indicate that irradiation of (*P,P*)-*trans* isomer at λ ≥ 280 nm at 60 °C will lead to the formation of all three other states of the four-step isomerization cycle. As such, the molecule will perform as a motor undergoing unidirectional rotation in a way schematically depicted in Figure 5.11. Relative to the bottom half of the molecule, which can be viewed as the static part (stator), the top half of the molecule (rotor) undergoes a full 360° rotation exclusively in a clockwise manner. The operation of this molecular motor relies on two photoisomerization and two thermal isomerization steps, which are controlled by the steric effects in the four isomeric structures.

A variety of molecular motors have been prepared and characterized based on crowded alkene systems after Feringa's report of the first photo-driven molecular motor [5]. These molecular motors have been subsequently applied as "engines" to power various molecular machines. The example illustrated in Figure 5.12 is a

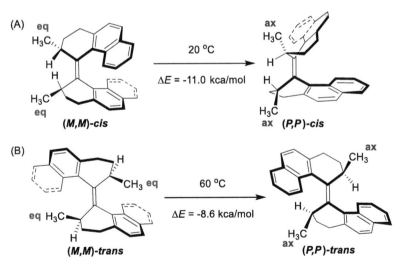

Figure 5.10 Thermal isomerization processes of (A) from (*M,M*)-*cis* to (*P,P*)-*cis* and (B) from (*M,M*)-*trans* to (*P,P*)-*trans*.

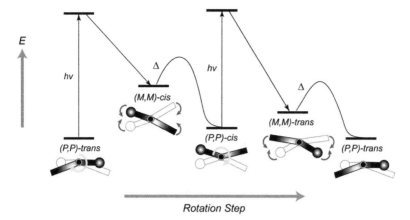

Figure 5.11 Schematic illustration of the operational sequence and potential energy profile of Feringa's first-generation molecular motor.

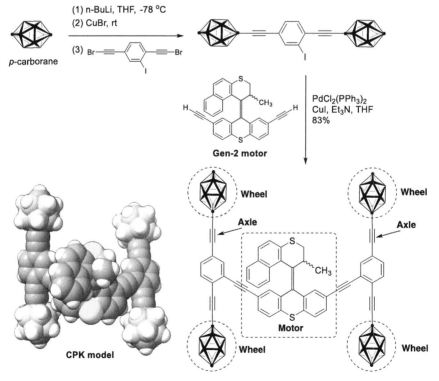

Figure 5.12 Construction of a *p*-carborane-wheeled molecular car that is equipped with a second-generation molecular motor.

molecular vehicle (termed a motorized nanocar) designed and synthesized by Tour and co-workers in 2006 [6]. In their work, four cage-shaped *p*-carborane groups were utilized to perform as the four wheels of a nanocar. The four *p*-carborane groups are connected to linear alkynyl groups, which serve as the axles. In the central region of this nanocar, a second-generation molecular motor developed by the Feringa group [7] is installed through a Sonogashira cross-coupling reaction. Kinetics studies in the solution phase showed that the molecular motor unit could rotate upon irradiation with light at 365 nm. However, the performance of this motorized nanocar on a surface has not yet been clearly demonstrated.

In 2016, Tour and co-workers designed a molecular machine, termed nanoroadster (Figure 5.13), which showed photoinduced translational motion on a surface [8]. The molecular structure of the nanoroadster consists of two adamantanyl wheels attached to alkynyl axles, to which a fast photo-driven crowded alkene-based molecular motor is connected. To test its performance as a feasible molecular vehicle, the nanoroadster was loaded on a Cu(111) surface and imaged by scanning tunnelling microscopic (STM) analysis to monitor the surface motion of this nanoroadster under various conditions. It was found that the thermal energy at a very low temperature (6 K) was insufficient to drive the motor in the nanoroadster to undergo the required helix inversion steps. Increasing the temperature to 150 K, diffusion of the molecule across the surface occurred without any irradiation. Under photoirradiation with laser lights (e.g., at 266 nm), the motor unit in the nanoroadster started to rotate, which in turn considerably increased the speed of diffusion in comparison to that of thermal diffusion. The direction of photoinduced motion of the nanoroadster was observed to be random, indicating that the performance of the nanoroadster as a molecular vehicle is not ideal. Nevertheless, this work has clearly demonstrated that a photo-driven molecular motor is capable of enhancing the movement of a molecular car.

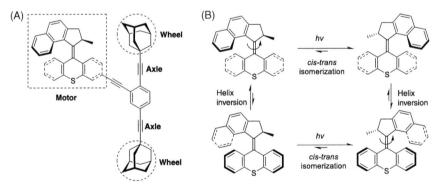

Figure 5.13 (A) Molecular structure of a nanoroadster designed by Tour and co-workers. (B) Rotatory mechanism for the photo-driven motor in this nanoroadster.

5.5 Case Study II: An Electrically Driven Molecular Car

It is reasonable to envision that a molecular motor(s) that rotates in a consecutive and unidirectional manner should propel a nanocar into directed motion. However, to achieve such a performance is very challenging, because this will require a well-designed molecule that can properly utilize light, chemical, and electrical energies to modulate its interactions with a surface. A *tour de force* in this research topic was achieved by Feringa and co-workers in 2011 [9]. As shown in Figure 5.14, they designed a four-wheel-drive nanocar, in which four crowded alkene moieties act as molecular motors that contain specific (R) and (S) configurations in a way that the entire molecule takes a *meso*-(R,S–R,S) structure. Individual molecules of this four-wheel-drive nanocar could be readily placed on a Cu(111) surface through sublimation and then imaged and investigated by STM.

(A)

(B)

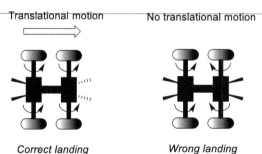

Translational motion   No translational motion

Correct landing   Wrong landing

Figure 5.14 (A) Structure of an electrically driven four-wheel-drive nanocar designed by Feringa and co-workers. (B) Orientations of the nanocar on a surface that enable or disable concerted wheel rotations for a translational motion.

Similar to the photo-driven molecular motors developed by the Feringa group, the molecular motors in this nanocar can operate to give unidirectional rotations through consecutive C=C double bond isomerization and helix thermal inversion steps (see Figure 5.15B). When loaded on a metal surface with a proper orientation (correct landing) as depicted in Figure 5.14, the four molecular motors (wheels) would rotate in a concerted manner, resulting in linear translational motion of the nanocar on a surface. Unlike the photo-driven motors mentioned in the previous subsection, the molecular motors in this four-wheel-drive nanocar are powered by electrical energy instead of photonic energy. The operation was achieved by placing an STM tip above the molecule with an applied voltage to electronically excited the molecule in order to induce the *cis–trans* isomerization of the C=C double bonds in the motors (Figure 5.15A). The detailed rotatory mechanism of the electrically driven motor is depicted in Figure 5.15C, in which four isomeric states are consecutively formed through two C=C *cis–trans* isomerization steps and two helix inversion steps. The four molecular motors could rotate in a concerted conrotary manner as shown in Figure 5.15D. It was observed through STM imaging that with 10 steps of excitation, the nanocar moved along a distance of 6 nm. When the nanocar landed on the surface in a different orientation (see the "wrong landing" drawing in Figure 5.14B), the combined effects of

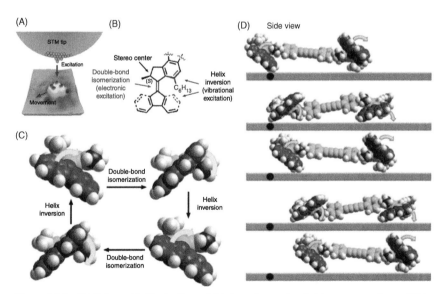

Figure 5.15 (A) Schematic illustration of driving a four-wheel-drive nanocar by an STM tip with a bias voltage. (B) Structural details of rotary units in the molecular motors. (C) Schematic illustration of the 360° rotation of the rotary motor. (D) Molecular model representation (side view) of the paddlewheel-like motion of the four-wheel-drive nanocar. (Reproduced from *Nature* 2011, *479*, 208–211).

the motor units cancelled out and the molecule was hence inhibited from translational motion. Experimentally, the wrongly landed nanocar was observed to lack
movement under the same electronic excitation conditions applied to the correctly landed one. Moreover, the (*R*,*R*–*R*,*R*) and (*S*,*S*–*S*,*S*) diastereomers of this
nanocar were synthesized and studied. The four wheels of the (*R*,*R*–*R*,*R*) and
(*S*,*S*–*S*,*S*) isomers were expected to rotate in a disrotatory manner and thus give a
circular motion. STM studies indicated that these two isomers move in a random
trajectory on a surface upon electronic excitation. Feringa's electrically driven
nanocar has clearly demonstrated that a single molecule equipped with four
molecular motors in a proper orientation can adsorb and convert external
electrical energy into unidirectional motion along a surface.

5.6 Case Study III: Chemically Driven Molecular Motors

Efforts of utilizing chemical fuels to drive molecular machines have been made by
synthetic chemists for decades. Figure 5.16 shows the first example of a chemically driven molecular motor developed by Kelly and co-workers in 1999 [10].
This motor was designed based on various molecular ratchets developed by the
Kelly group. Its molecular structure contains two important parts, a triptycene
"wheel" and a helicene "brake," which are connected through a C–C single bond.
The helicene unit interacts with the triptycene to raise the rotational energy

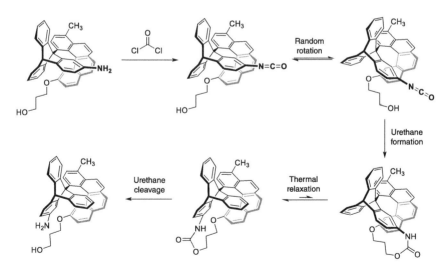

Figure 5.16 A chemically driven triptycene–helicene molecular motor developed by
Kelly and co-workers.

barrier for the C–C bond to 25 kcal mol^{-1}. As such, the rotation of the triptycene wheel relative to the helicene brake is considerably hindered. When the triptycene-helicene system is unsubstituted, the rate of rotation of the C–C bond is equal in both directions as predicted by the Second Law of Thermodynamics. The triptycene-helicene molecular rachet, even though showing a desymmetrized energy profile for rotation, cannot exhibit any unidirectional rotation about the C–C bond. To achieve motor performance (i.e., unidirectional rotation of the rotor), the triptycene-helicene structure was functionalized with a primary alcohol linked to the helicene, and an amino group on the triptycene wheel. As shown in Figure 5.16, treatment of this molecule with phosgene leads to the formation of an isocyanate group on the triptycene moiety. Thermal energy then drives the wheel to rotate clockwise, bringing the isocyanate close to the primary alcohol group to undergo a urethane forming reaction. The urethane intermediate then undergoes a unidirectional rotation to release the steric encumbrance, resulting in a further clockwise rotation. Finally, the urethane linkage is cleaved, and the product shows a single, nonrepeatable 120° rotation of the triptycene wheel relative to the helicene brake through the chemical and thermal steps mentioned above. Although the Kelly's molecular motor showed only a limited degree of unidirectional rotation, its design and testing represent a milestone in the research of molecular motors fueled by chemical energy.

In 2005, the first chemically driven molecular motor that is capable of 360° unidirectional rotation was achieved by Feringa and co-workers [11]. Their molecular motor is based on a 1-phenylnaphthalene system, in which a phenyl and a naphthyl group are connected through a single bond. The two aryl units act as the rotor and the stator, respectively. Figure 5.17 illustrates the operational sequence of this chemically driven molecular motor. The motor works through a power stroke mechanism, which is dependent on chemical reactivity to achieve unidirectional motion. The rotational sequence proceeds through four distinct stations, indicated as stations A to D in Figure 5.17. In stations A and C, the upper rotor and the lower stator units are interlocked by a lactone moiety. Stations B and D are unlocked structures, where the lactone tether is cleaved. At these unlocked stages, the rotation about the C–C bond between the two aryl groups is still restricted due to the steric interactions among their *ortho*-substituents. Such steric effects prevent the two unlocked structures from undergoing helix inversion, which is essential for the motor function.

The detailed chemical reaction steps that lead to a unidirectional rotation of this chemically driven molecular motor are described in Figure 5.18. Starting from station A, a stereoselective reduction reaction was conducted using BH$_3$ in combination with (*S*)-2-methyl-oxazaborolidine, which is a chiral catalyst well known as the Corey–Bakshi–Shibata catalyst or the CBS catalyst. The reduction reaction converted the lactone moiety into a phenol and a primary alcohol group, resulting in an unlocked phenylnaphthalene. The role of the (*S*)-CBS catalyst in

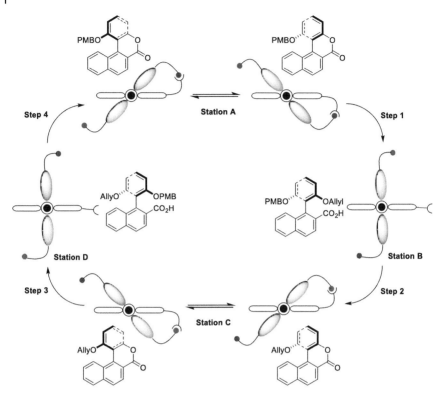

Figure 5.17 Schematic illustration of the operation of the first chemically driven molecular motor developed by Feringa and co-workers.

this reaction is critical, since it promotes a clockwise rotation of the rotor, resulting in excellent enantioselectivity. The phenol group in the reduced product was next protected as an allyl ether. After this transformation, the primary alcohol on the naphthyl group was oxidized into an aldehyde by CrO_3 and then into a carboxyl group by $NaClO_2$ in the presence of 2-methyl-2-butene, arriving at station B in a good yield and accomplishing step 1 as illustrated in Figure 5.17. Overall, the unidirectional rotation of step 1 can be viewed as being induced by the chirality of the fuel (i.e., BH_3 and (S)-CBS) used for breaking the lactone bond.

In the structure of station B, the two phenol groups are orthogonally protected by p-methoxybenzyl (PMB) and allyl groups, respectively. To continue the clockwise rotation, in step 2, the PMB group was selectively removed by $Ce(OTf)_3$ in the presence of dimethoxybenzene to yield a free phenol intermediate, which spontaneously cyclized into a lactone to afford station C. Step 3 underwent chemical transformations similar to those in step 1. Herein, an asymmetric reduction of the lactone group was carried out and the resulting phenol group was then protected

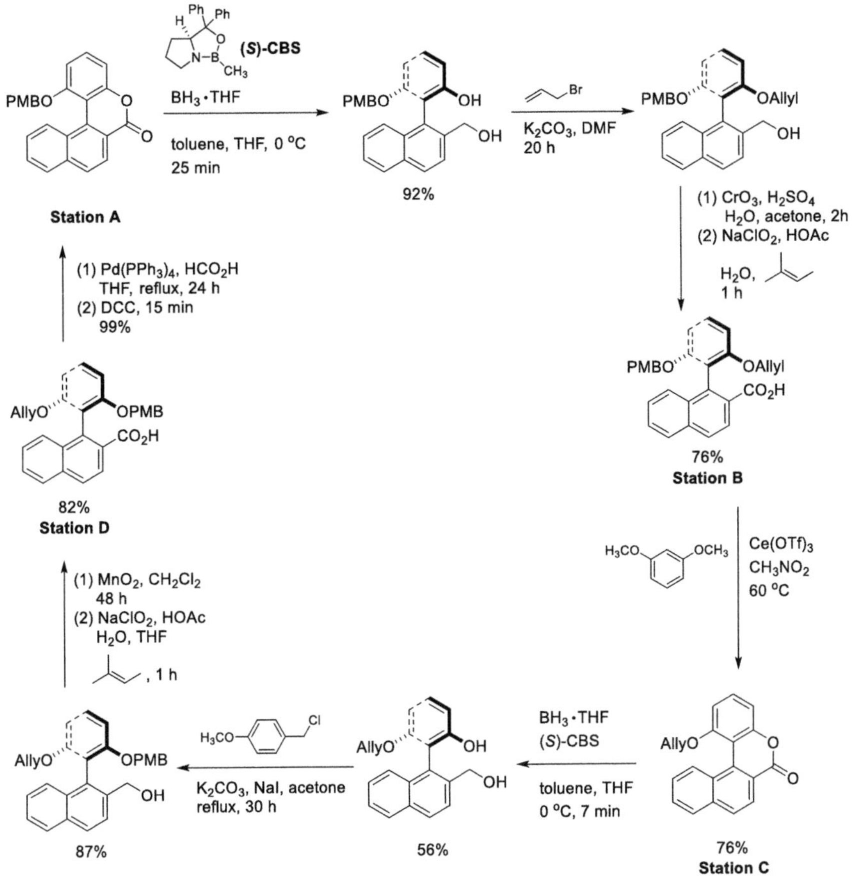

Figure 5.18 Detailed chemical transformations involved in the 360° unidirectional rotation of Feringa's chemically driven molecular motor.

as a PMB ether. Again, the use of (S)-CBS catalyst promoted a clockwise axle rotation between the rotor and stator units. After that, the primary alcohol group was sequentially oxidized into a carboxyl group, yielding the unlocked station D. Finally, in step 4, the allyl ether group was selectively cleaved with Pd(PPh₃)₄ and formic acid. The resulting phenol group further reacted with the carboxyl unit through a DCC promoted lactonization reaction to return to the structure of station A. Likewise, step 4 is highly enantioselective, the occurrence of which eventually fulfilled a 360° unidirectional rotation of the upper phenyl rotor with respect to the lower naphthalene stator. Overall, the chemical reactions and purification steps employed in operating this chemically driven molecular motor are very complex. Practically speaking, this chemically driven molecular motor is less useful than photo-driven molecular motors. Nevertheless, the accomplishment of the controlled unidirectional rotation of this motor provides a valuable example for understanding how chemically driven molecular machines work at the molecular level.

5.7 Case Study IV: A Redox-driven Molecular Pump

A large number of synthetic molecular machines have been prepared and tested in the solution phase to prove their working principles as well as applicability. On the other hand, precisely controlled operation of molecular machines in solution is extremely challenging, since molecules are subjected to random thermal fluctuations or Brownian motions as well as constant bombardment by solute molecules. In the biological world, protein-based molecular machines use external stimuli (e.g., chemical inputs) to achieve specific and directed movements and thus can produce useful work. Electron transfer processes are of crucial importance in many biological processes, regulating both energy transduction and signal transmission. At the molecular level, electron transfers result in redox reactions that may switch a molecular system among multiple oxidation states. In contrast to photochemical reactions, redox reactions do not yield any photostationary states. They can proceed reversibly in both directions of a reaction through controlled oxidation and reduction steps. Moreover, redox-regulated molecular systems are more compatible with biological systems than photoactivated ones. Synthetically, redox-active units can be readily integrated in a multicomponent molecular system, in which redox processes can be executed through numerous approaches, including addition of chemical oxidants/reductants to a solution, photoinduced electron transfer, and heterogeneous electron transfer reactions by means of applying appropriate electrical potentials to conducting electrodes. The majority of redox-driven synthetic molecular machines were designed based on catenane and rotaxane interlocked molecular structures, in which redox reactions can be applied to control either the coordination geometry around transition metal cations or the electron donor-acceptor interactions between different molecular components.

Figure 5.19A illustrates the structure of an artificial molecular pump, namely **DB**·3PF$_6$, which was developed by Stoddart and co-workers in 2015 [12]. The design of this molecular pump was inspired by the functions of certain biochemical machinery in living organisms. For example, carrier proteins embedded in a cell membrane can bind specific solute molecules and then undergo a series of conformational changes to transfer them across the membrane. In a sense, this type of protein machine is fuelled by chemical energy to perform the thermodynamically uphill work of moving solute molecules or ions against a concentration gradient, transporting them from low concentration to high concentration regions.

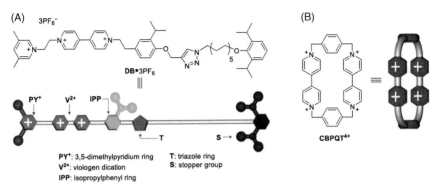

Figure 5.19 (A) Molecular structure and graphical representation of an artificial molecular pump (**DB**·3PF$_6$) designed by Stoddart and co-workers. (B) Molecular structure and graphical representation of **CBPQT**$^{4+}$. (Reproduced from *Nat. Nanotechnol.* 2015, *10*, 547–553 with modifications).

Structurally, the molecular pump **DB**·3PF$_6$ consists of multiple components covalently linked in a molecular dumbbell shape. It was designed to be threaded through by a tetracationic macrocycle, cyclobis(paraquat-*p*-phenylene) (**CBPQT**$^{4+}$) (Figure 5.19B), to form a mechanically interlocked molecular structure called a rotaxane. In Chapter 1, we have discussed that **CBPQT**$^{4+}$ shows charge-transfer (CT) interactions with π-conjugated organic donors. Utilizing this type of non-covalent force, **CBPQT**$^{4+}$ can be kinetically trapped in various molecular systems to form catenanes and rotaxanes. In the structure of **DB**·3PF$_6$, one terminal position is end capped with a 3,5-dimethylpyridinium group denoted as **PY**$^+$, and the other terminus is a bulky 2,6-diisopropylphenoxy group denoted as **S**. In the central region of the molecular dumbbell are a dicationic viologen unit (**V**$^{2+}$), an isopropylphenyl ring (**IPP**), and a 1,2,3-triazole unit (**T**). All these components are linked together through flexible alkyl chains, and they work synergistically to achieve the function of driving **CBPQT**$^{4+}$ rings away from equilibrium toward a higher local concentration; that is, from the solution phase to a ring-collection region of the molecular dumbbell.

Before examining the operational details of the molecular pump, let's first go through the synthesis of this compound. Stoddart and co-workers used a convergent synthetic approach to assemble this molecular pump (see Figure 5.20). Two chain-like structures that carry a terminal alkyne and a terminal azido group, respectively, were first generated through multiple synthetic steps. Each of the chains contains the components designed for the molecular pump. These two precursors were eventually subjected to an efficient CuAAC (click) reaction in the presence of a Cu(I) catalyst, Cu(CN)$_4$PF$_6$, and tris((1-benzyl-4-triazolyl)methyl) amine (**TBAT**) which is a ligand commonly used to enhance the catalytic effect of the copper catalyst used in click chemistry. The relatively high yield of the final

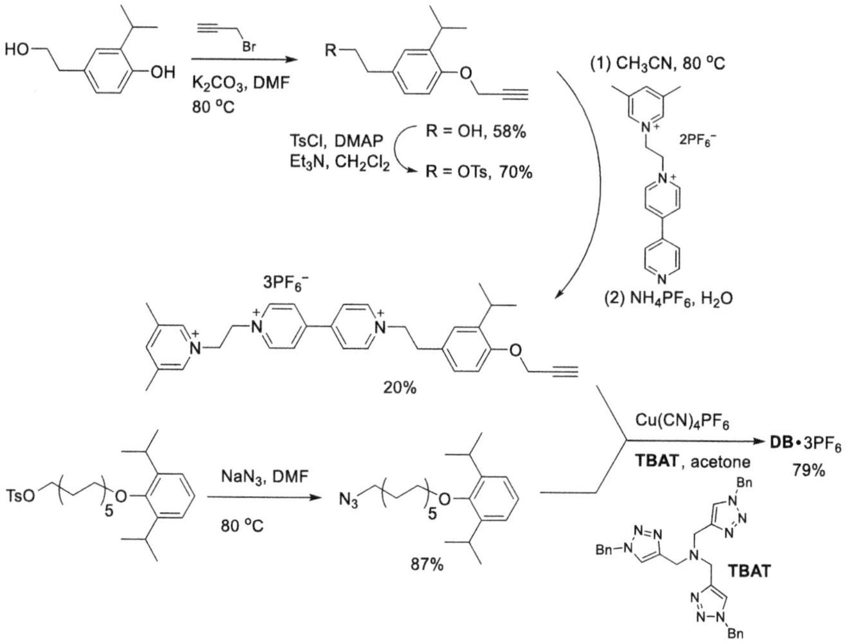

Figure 5.20 Synthesis of **DB**·3PF$_6$ through a divergent approach.

product testifies to the usefulness of click chemistry in the preparation of nano-materials and devices.

The operation of the molecular pump was accomplished in the solution phase by mixing **DB**·3PF$_6$ with tetracationic cyclobis(paraquat-*p*-phenylene) (**CBPQT**$^{4+}$). The detailed steps of the operation are schematically illustrated in Figure 5.21. At the initial stage (Figure 5.21A), the **CBPQT**$^{4+}$ and **DB**·3PF$_6$ are both positively charged and therefore subjected to repulsive Coulombic forces. The trapping of a **CBPQT**$^{4+}$ ring by **DB**·3PF$_6$ is therefore thermodynamically dis-favored. Upon addition of Zn dust, the **CBPQT**$^{4+}$ unit can be reduced into **CBPQT**$^{2(+•)}$ and **V**$^{2+}$ to **V**$^{+•}$. The reduction reactions attenuate the Coulombic repulsion between the **CBPQT**$^{2(+•)}$ ring and the **PY**$^+$ end group, thus allowing **CBPQT**$^{2(+•)}$ to pass through the **PY**$^+$ barrier and thread onto the molecular dumb-bell to form a thermodynamically stable rotaxane. This step is described schemat-ically in Figure 5.21B, in which the **CBPQT**$^{2(+•)}$ ring is parked at the **V**$^{+•}$ unit, forming a trisradical tricationic complex **V**$^{+•}$⊂**CBPQT**$^{2(+•)}$ through favored π–π interactions. In the next step, the rotaxane is oxidized with NOPF$_6$, restoring the redox-active components to their fully charged states. As such, the **CBPQT**$^{4+}$ and **V**$^{2+}$ units are pushed away from each other due to their significant Coulombic repulsion (Figure 5.21C). Now, if the **CBPQT**$^{4+}$ ring moves back to pass through

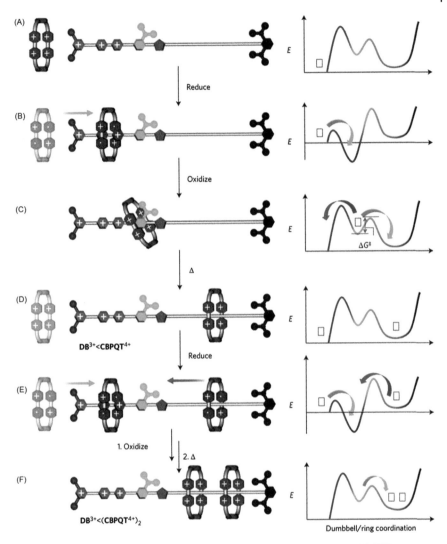

Figure 5.21 Schematic illustration of the operating mechanism for the **DB**·3PF$_6$ molecular pump used to trap two **CBPQT**$^{4+}$ rings through two redox cycles. (Reproduced from *Nat. Nanotechnol.* 2015, *10*, 547–553 with modifications).

the **PY**$^+$ ring, the rotaxane will be disassembled. Although this is a thermodynamically favored pathway, kinetically it is prohibited. As illustrated by the energy profile shown in Figure 5.21C, this process needs to overcome a higher energy barrier due to the repulsive Coulombic forces between **CBPQT**$^{4+}$ and **PY**$^+$. Instead, movement of the **CBPQT**$^{4+}$ ring further along the molecular dumbbell to overcome the **IPP** barrier is much easier. Eventually, the molecular pump undergoes a sequence of co-conformational rearrangements, driving the **CBPQT^{4+}** ring to traverse the **IPP** unit, trapping it on the oligomethylene chain which is a ring-collection region (Figure 5.21D). The first sequence of reduction, oxidation, and co-conformational rearrangements serves as a flash-energy ratchet mechanism to pump one **CBPQT^{4+}** ring from the solution to the ring-collection region of the molecular dumbbell.

After the first redox cycle, the molecular pump can continue to trap a second **CBPQT^{4+}** ring using the same conditions. However, the second ring trapping faces two challenging issues: (1) reverse transport of the first **CBPQT^{4+}** ring needs to be avoided, and (2) the second **CBPQT^{4+}**ring has to be driven into proximity with the first one, in which significant Coulombic repulsion cannot be avoided. To address the first challenge, the energy profile should take a shape as shown in Figure 5.21E, in which the trapping of the second **CBPQT$^{2(+\cdot)}$** ring by the **V$^{+\cdot}$** unit is kinetically faster than the crossing of the first **CBPQT$^{2(+\cdot)}$** ring through the **IPP** unit. Stoddart and co-workers have demonstrated that their elegantly designed molecular pump indeed features such an energy profile. The **IPP** unit imposes sufficient steric hindrance to prevent the first **CBPQT$^{2(+\cdot)}$** ring from moving back at the stage of the second reduction step (Figure 5.21). As for the second challenge, the trapping of two **CBPQT^{4+}** rings was successfully achieved in Stoddart's work and they further proposed that if the length of the ring-collection oligomethylene unit could be further elongated, additional **CBPQT^{4+}** rings could possibly be pumped into this region through multiple redox cycles.

Further Reading

- Balzani, V.; Venturi, M.; Credi, A., Molecular Devices and Machines: A Journey into the Nanoworld. 2nd ed.; Wiley-VCH: Weinheim, 2008.
- Coyle, J. D., Introduction to Organic Photochemistry. John Wiley & Sons: New York, 1991.

References

1 Allison, W. S., F1-ATPase: A Molecular Motor that Hydrolyzes ATP with Sequential Opening and Closing of Catalytic Sites Coupled to Rotation of Its γ Subunit. *Acc. Chem. Res.* **1998**, *31*, 819–826.
2 (a) Erbas-Cakmak, S.; Leigh, D. A.; McTernan, C. T.; Nussbaumer, A. L., Artificial Molecular Machines. *Chem. Rev.* **2015**, *115*, 10081–10206; (b) Coskun, A.; Banaszak, M.; Astumian, R. D.; Stoddart, J. F.; Grzybowski, B. A., Great Expectations: Can Artificial Molecular Machines Deliver on Their Promise?

Chem. Soc. Rev. **2012**, *41*, 19–30; (c) Lancia, F.; Ryabchun, A.; Katsonis, N., Life-Like Motion Driven by Artificial Molecular Machines. *Nat. Rev. Chem.* **2019**, *3*, 536–551; (d) Krause, S.; Feringa, B. L., Towards Artificial Molecular Factories from Framework-Embedded Molecular Machines. *Nat. Rev. Chem.* **2020**, *4* (10), 550–562.

3 Harada, N.; Koumura, N.; Feringa, B. L., Chemistry of Unique Chiral Olefins. 3. Synthesis and Absolute Stereochemistry of *trans*-and *cis*-1,1′,2,2′,3,3′,4,4′-Octahydro-3,3′-dimethyl-4,4′-biphenanthrylidenes. *J. Am. Chem. Soc.* **1997**, *119*, 7256-7264.

4 Koumura, N.; Zijlstra, R. W.; van Delden, R. A.; Harada, N.; Feringa, B. L., Light-Driven Monodirectional Molecular Rotor. *Nature* **1999**, *401*, 152–155.

5 (a) Kassem, S.; van Leeuwen, T.; Lubbe, A. S.; Wilson, M. R.; Feringa, B. L.; Leigh, D. A., Artificial Molecular Motors. *Chem. Soc. Rev.* **2017**, *46*, 2592–2621; (b) Baroncini, M.; Silvi, S.; Credi, A., Photo-and Redox-Driven Artificial Molecular Motors. *Chem. Rev.* **2019**, *120*, 200–268.

6 Morin, J.-F.; Shirai, Y.; Tour, J. M., En Route to a Motorized Nanocar. *Org. Lett.*. **2006**, *8*, 1713–1716.

7 Koumura, N.; Geertsema, E. M.; van Gelder, M. B.; Meetsma, A.; Feringa, B. L., Second Generation Light-Driven Molecular Motors. Unidirectional Rotation Controlled by a Single Stereogenic Center with Near-Perfect Photoequilibria and Acceleration of the Speed of Rotation by Structural Modification. *J. Am. Chem. Soc.* **2002**, *124*, 5037–5051.

8 Saywell, A.; Bakker, A.; Mielke, J.; Kumagai, T.; Wolf, M.; García-López, V.; Chiang, P.-T.; Tour, J. M.; Grill, L., Light-Induced Translation of Motorized Molecules on a Surface. *ACS Nano* **2016**, *10*, 10945–10952.

9 Kudernac, T.; Ruangsupapichat, N.; Parschau, M.; Maciá, B.; Katsonis, N.; Harutyunyan, S. R.; Ernst, K.-H.; Feringa, B. L., Electrically Driven Directional Motion of a Four-Wheeled Molecule on a Metal Surface. *Nature* **2011**, *479*, 208–211.

10 Kelly, T. R.; De Silva, H.; Silva, R. A., Unidirectional Rotary Motion in a Molecular System. *Nature* **1999**, *401*, 150–152.

11 Fletcher, S. P.; Dumur, F.; Pollard, M. M.; Feringa, B. L., A Reversible, Unidirectional Molecular Rotary Motor Driven by Chemical Energy. *Science* **2005**, *310*, 80–82.

12 Cheng, C.; McGonigal, P. R.; Schneebeli, S. T.; Li, H.; Vermeulen, N. A.; Ke, C.; Stoddart, J. F., An Artificial Molecular Pump. *Nat. Nanotechnol.* **2015**, *10*, 547–553.

6

Computational Modeling and Simulations in Organic Nanochemistry

6.1 Introduction to Computational Chemistry

Organic chemistry is a branch of science that is deeply rooted in experiments and analyses. In the history of organic chemistry, extensive experimental investigations have been conducted to gain understanding of the properties of organic compounds and reactions. Experiments have in turn led to observations and measurements, allowing researchers to know the outcomes of various chemical processes or phenomena. Analyses, on the other hand, spurred the development of hypotheses, models, and theories to interpret the experimental outcomes. As a matter of fact, observation and interpretation are the twin pillars of the fundamental approach for all scientific inquiries; the first one tells us how experimental results occur, while the second leads us to understand why such experimental outcomes occur. Based on the knowledge gathered from experiments and analyses, researchers are able to make reasonable prediction of outcomes for unknown systems and carry out rational experimental designs in their future work.

The field of chemistry began when humans started to experiment with the manipulation of known substances. Laboratorial experiments including various tests and trials are common practices used to acquire understanding on an empirical basis. With empirical experimental results in hand, conceptual frameworks and theories were developed to rationalize what has been observed and to provide guidance to the next round of experimental explorations. The latter step is known as theorization, which is another cornerstone of scientific inquiries. Theorization can be achieved by using models, which is a way of creating simplified representations of complex phenomena that would not be readily comprehended otherwise. For example, the concept of valency (or valency number) in chemistry was first proposed by Edward Franklin in 1852, who observed that certain elements show the tendency to combine with other elements to form compounds that have equivalent numbers of groups (e.g., NH_3, NO_3, NCl_3). The

Organic Nanochemistry: From Fundamental Concepts to Experimental Practice,
First Edition. Yuming Zhao.

"combining power" of an element was recognized as **valency** or **valence** (from Latin *valentia*, meaning strength or capacity). This conceptualization has not only set the foundation for modern chemical bonding theories, but later nurtured the development of other important concepts and theories on chemical bonds; for example, Lewis structures, the hybridization model, VSEPR theory, and the valence bond (VB) model. Intuitive models and pertinent theories have greatly enhanced the understanding of fundamental molecular properties. For instance, Chapter 1 has introduced how to use the hybridization model and VSEPR theory to quickly determine covalent bond types as well as to predict molecular geometries. In dealing with the class of aromatic compounds, the Hückel's (4n + 2) rule serves as an effective tool to assign aromaticity or anti-aromaticity for various cyclic and polycyclic π-conjugated systems. With such notions, the chemical stability, reactivity, and magnetic properties of a wide range of organic compounds can now be qualitatively predicted and compared in a straightforward and intuitive manner.

In the field of organic synthesis, many reaction models have been developed to interpret the complex reactivities observed from synthetic experiments. Let's take the nucleophilic addition of an acyclic carbonyl compound as an example. When a carbonyl compound has a stereogenic center at the α-position, the addition reaction occurring on it becomes stereoselective. In 1952, Donald J. Cram proposed a simple rule to rationalize and predict the stereochemistry outcomes for this type of reaction [1]. As illustrated in 6.1A, the **Cram's rule** of asymmetric induction is set upon the argument of steric bias around the C=O group of the reactant. The Cram model stipulates that the carbonyl group is gauche to the medium- and small-sized groups on the adjacent carbon. The favored direction of nucleophilic attack occurs on the less hindered (i.e., small group) side. While this rule has been found practically useful in explaining the outcomes of a large number of reactions, concerns about the theoretical soundness of Cram's model were raised as well. One argument was that Cram's model does not represent a reasonable transition state for carbonyl addition. Numerous revised models ensued to provide theoretically more reasonable explanations. Among them, the **Felkin-Ahn model** [2] is a widely accepted one by the organic chemistry community. This model is rooted upon the experimental observation that a nucleophile always favors attacking the carbonyl group along an angle of about 107°. Such an angle is well known as the **Bürgi-Dunitz angle**, which is named after two crystallographers, H. B. Bürgi and J. D. Dunitz. They deduced a general reaction path for the nucleophilic attack on carbonyl through examination of the intramolecular N···C=O interactions in the single crystal structures of a series of organic compounds, each of which contains a carbonyl and a tertiary amino group [3]. Following this trajectory, sterically favored transition states can be conceived as shown in Figure 6.1B. Although the Felkin-Ahn model predicts the same stereochemical outcomes as Cram's

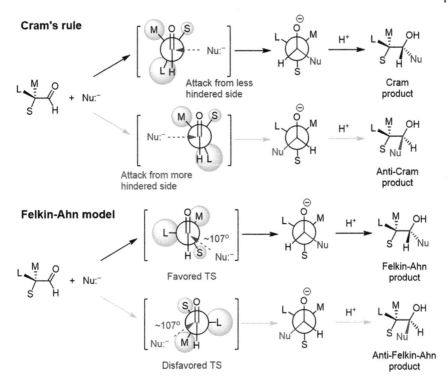

Figure 6.1 Two empirical models, Cram and Felkin-Ahn, used to explain the stereoselective nucleophilic attack on a carbonyl compound with an α-stereogenic center. The letters L, M, and S denote large-, medium-, and small-sized substituent groups, respectively.

model does, it makes a better theoretical sense in terms of transition state favored selectivity. In a later section of this chapter, we will illustrate the search of a Felkin-Ahn-type transition state through quantum chemistry calculations.

The evolution of chemical models is not just limited to the conceptual and theoretical frameworks, but often involves complex mathematical equations and computational methods. The establishment of the theory of quantum mechanics (QM) in the 1920s opened a new approach to describe the physical properties of nature at the atomic and subatomic levels. The hallmark of this field is the theory based on pure mathematics that explains how chemistry arises from the interactions of nuclei and electrons. Heisenberg, Schrödinger, Dirac, and other prominent physicists made ground-breaking contributions in the early 1920s by developing the wave mechanics in quantum theory. The introduction of computer technology to chemical research in the 1950s marked the inception of modern computational chemistry. Over the past decades, computational chemistry has made great impacts in the advancement of modern organic chemistry. In today's research, organic chemists not only deal with synthesis and characterization of organic compounds in the lab, but also perform *in silico* designs of functional organic materials and computational modeling of sophisticated organic reaction processes. The rise of artificial intelligence (AI) technology in the past decade has brought a revolutionary transformation upon the landscape of science and engineering. In this era of data science, even the field of organic chemistry has not escaped its profound impacts. Different from the previous chapters, this chapter focuses on introducing the basic knowledge of computational organic chemistry and providing a glimpse into a variety of computational tools useful for enriching research in organic nanochemistry.

6.2 Computational Methods for Modeling from Macroscopic to Nanoscale Systems

Materials ranging from bulk engineering components to small molecular clusters are dramatically different in size and time scales required for analysis. Computational methods for simulating their properties and performances are therefore fundamentally different (see Figure 6.2). For example, to design a ship propeller, engineers would assume that all the substances of objects (metal, fluid, air, etc.) completely fill the spaces they occupy, which is known as the continuum mechanics (CM) approach. Detailed solutions to structural deformation, mechanical properties, and fluidic responses can be obtained by using popular methods in the fields of engineering and physics, such as finite element analysis (FEA) and computational fluidic dynamics (CFD).

The mechanics to be solved heavily rely on Newton's three laws of motion. The first law stipulates that an object at rest tends to stay at rest, while an object in motion tends to stay in motion with the same direction and speed. The tendency to resist changes in a state of motion is called **inertia**. Newton's third law of motion says that for every action (force) there is an equal and opposite reaction (force). Newton's second law can be expressed through Eq. 6.1, which describes the relationship between time and the evolution of the coordinates and velocity of an object.

$$ma = F \tag{6.1}$$

In this equation m is the mass of the particle, a is its acceleration, and F is the instantaneous force applied on the particle. The acceleration a in a 1D motion can be further written as,

$$a = \frac{dv}{dt} = \frac{dx^2}{dt^2} \tag{6.2}$$

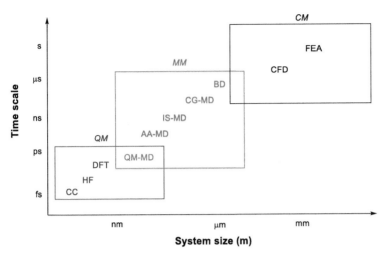

Figure 6.2 Simulation theories suited for different scales of time and system size: quantum mechanics (QM) including coupled cluster (CC), Hartree-Fock (HF), and density functional theory (DFT) methods; molecular mechanics (MM) including all-atom molecular dynamics (AA-MD) simulations, implicit solvent and coarse-grained MD (IS-MD and CG-MD), and Brownian dynamics (BD); continuum mechanics (CM) including computational fluidic dynamics (CFD) and finite element analysis (FEA). The ranges of time and sizes are approximate.

where v is the velocity of the particle, x is the position of the particle, and t is the time. The force F can be related to the potential energy $V(x)$ by,

$$F = -\frac{dV(x)}{dx} \tag{6.3}$$

Combining the above equations, one can rewrite Newton's second law as a second-order differential equation that relates the evolution of coordinate x with time t

$$\frac{d^2x}{dt^2} = -\frac{1}{m}\frac{dV(x)}{dx}. \tag{6.4}$$

For a particle in 3D motion, the acceleration a is related to the location of the particle as follows:

$$a = \frac{dv}{dt} = \frac{d^2r}{dt^2}. \tag{6.5}$$

In three dimensions, the location of a particle is specified by its location vector, $r = ix + jy + kz$, where i, j, and k define the vector for the location of r in reference to the origin of a Cartesian coordinate system. The force F can now be written as

$$F = \nabla V(r) \tag{6.6}$$

where the symbol ∇ is used to represent a first-order gradient operator shown in Eq. 6.7

$$\nabla = i\frac{d}{dx} + j\frac{d}{dy} + k\frac{d}{dz}. \tag{6.7}$$

For a single particle moving in three dimensions, the sum of kinetic and potential energy is defined by a Hamiltonian function H,

$$H = \frac{p^2}{2m} + V(r) \tag{6.8}$$

where $p^2/2m$ is the kinetic energy and $V(r)$ is the potential energy.

When the size and time scales become much smaller, for example, at the levels of proteins and organic molecules, the laws that govern the mechanics and dynamics are shifted to the theories built on quantum mechanics (QM) and molecular mechanics (MM). Depending on the scales and computational expenses, different methods can be adopted to acquire solutions. The QM-based methods such as cluster-coupled (CC), Hartree-Fock (HF), and density functional theory (DFT) are particularly useful in solving the electronic and energetic properties of molecular systems. A central task in QM-based calculations is to solve the **Schrödinger equation**, which is named after Erwin Schrödinger who postulated the equation in 1925. This ground-breaking wave equation changed the face of quantum theory and led Schrödinger to winning the Nobel Prize in Physics in 1933.

In the QM theory, a single particle (e.g., electron, nucleus) is proposed to be associated with a wavefunction $\psi(x,y,z,t)$, where x,y,z are the Cartesian coordinates that define the position of the particle and t is the time. For a many-particle system, ψ becomes a function containing the coordinates of all the particles. The volume integration of the wavefunction, $\int \psi^2 d\tau$ is the probability of finding these particles in the integrated space at time t.

Now, we come to another hypothesis of the QM theory. For every observable property, there is a linear Hermitian operator that is obtained by expressing the classical form in x,y,z Cartesian coordinates and momenta p,

$$p = \frac{\hbar}{i}\left(\frac{\partial}{\partial x} + \frac{\partial}{\partial y} + \frac{\partial}{\partial z}\right) \tag{6.9}$$

where $\hbar$ is called the reduced Planck constant ($\hbar = \dfrac{h}{2\pi} = 1.055 \times 10^{-34} J\ s^{-1}$).

With this hypothesis, the time-dependent Schrödinger equation is established as Eq. 6.10, where H is the Hamiltonian operator

$$i\hbar \frac{\partial}{\partial t}\Psi = H\Psi. \tag{6.10}$$

The time-dependent Schrödinger equation can be solved through the method of separation of variables, in which the wavefunction is written as a product of a temporal part and a spatial part

$$\psi(x_1, y_1, z_1, x_2, y_2, z_2, \ldots t) = \psi(x_1, y_1, z_1, x_2, y_2, z_2, \ldots)f(t). \tag{6.11}$$

By this method, the Schrödinger equation can now be reduced to a time-independent form,

$$H\psi = E\psi \tag{6.12}$$

where H is the Hamiltonian operator, ψ is the time-independent wavefunction, and E is a constant. Note that Eq. 6.12 is an **eigenvalue equation**, meaning that an operator acts upon a function to produce a multiple of the function itself as its results. E in this equation is called an **eigenvalue**. Solving of the time-independent Schrödinger equation yields a series of discrete E values, which are the energies corresponding to different stationary states of the wavefunction.

In practice, solving the time-independent Schrödinger equation is very challenging and requires several approximations. The first one is the **Born-Oppenheimer (BO) approximation**, which considers that the masses of nuclei in molecules are much greater (thousands of times) than that of an electron. The nuclei therefore move very slowly with respect to electrons, and the electrons react instantaneously to the changes in nuclear position. This approximation allows the motions of nuclei and electrons to be separated. Essentially, solving the Schrödinger equation can be simplified as solving an electronic Schrödinger equation, in which the nuclear positions are fixed as parameters. The electronic Schrödinger equation is expressed as

$$H_{el}\psi = E_{el}\psi \tag{6.13}$$

$$H_{el} = -\sum_i^N \frac{1}{2}\nabla_i^2 + \sum_{i \neq i'} V(r_i - r_{i'}) + \sum_{i,j} V(r_i - R_j) \tag{6.14}$$

where H_{el} is the Hamiltonian that contains the terms of electronic kinetic energy, electron-electron repulsion, and electron-nucleus attraction, respectively. Through this equation, the total electronic energy of a molecular system can be related to the spatial configuration of the nuclei in a molecule (i.e., nuclear geometry). A plot of the total energy as a function of the geometry of the nuclei produces a **potential energy surface** (PES), which is a very important concept for understanding various molecular structures and chemical reactivities in the ground and excited states.

It is also worth noting that the nuclei of molecules are not stationary. At a given temperature, the nuclei of a molecule fluctuate on a PES and such changes can be simulated to provide insights into relevant physical and chemical behaviors, for example, conformational variations, configurational isomerization, and chemical reaction paths. Most of the PESs are multidimensional and cannot be directly visualized since a human's visualization ability is confined within the 3D regime. For a pedagogical purpose, a 3D surface is commonly used to illustrate the key features on various PESs. Figure 6.3 presents a PES that correlates two molecular geometric parameters (R_1 and R_2) with the total electronic energy (E). On this surface there exist several stationary points, which are energy minima and maxima, respectively. Supposing the surface is a mountainous area, and there is a hiker exploring within it. When the hiker arrives at a valley, it is the lowest position as the hiker looks at all directions from it. The valley is therefore called an **energy minimum**. If the hiker travels from one valley to another valley, many possible routes exist but the hiker chooses only the most potential energy-saving one. Walking along such a path, the hiker needs to traverse a highest position

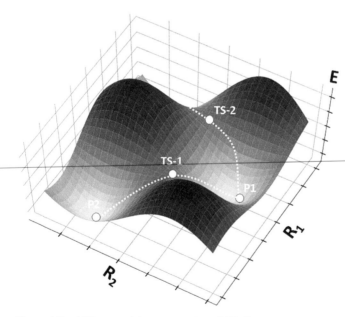

Figure 6.3 A 3D potential energy surface (PES) illustrating the positions of two transition states (**TS-1** and **TS-2**) and two energy minima (**P1** and **P2**). The dashed lines describe the trajectories of the transformations among the minimum-energy structures on the PES.

called a **transition state** (TS). A transition state is a saddle point, meaning that when arriving at this spot the hiker stands on the highest position along the direction of the hiking trail but remains the lowest position in all other directions. In Figure 6.3, there are two valleys **P1** and **P2**, representing two stable molecular structures. The transformation from **P1** into **P2** proceeds through a reaction trajectory highlighted by a dashed line. The transition state **TS-1** on this pathway is the highest-energy state for the reaction to overcome. The energy difference between **TS-1** and **P1** is called the **activation energy barrier** for the **P1**→**P2** transformation. It is worth mentioning that the transformation from **P2** to **P1** follows the exact same reaction trajectory. In other words, these two processes go through the same transition state **TS-1**. This is known as the **Principle of Microscopic Reversibility**, which states that any forward reaction and its reverse reaction should follow the exact same reaction trajectory. If the transition state of a forward reaction is found, then it is the same transition state as the reverse reaction.

As illustrated in Figure 6.3, an energy minimum represents a stable molecular structure, and the frequencies of its molecular vibrational modes should all be positive values. Let's consider molecular vibrations using the simple ball-and-spring model, where chemical bonds are treated as typical springs with positive force constants. The stretching and bending of a spring away from its equilibrium state consumes energy. In an energy minimum, all vibrational modes occur in association with positive force constants (normal bonds). Therefore, their frequencies are positive numbers. At a transition state, however, the molecular vibration along the reaction trajectory is energy-releasing since it starts from an energy maximum in this direction. As a result, the vibrational frequency along this direction is an imaginary number rather than a positive real number. The presence of only one imaginary vibrational frequency is a key criterion for validating whether a stationary point located on a PES is a transition state or not.

6.3 Computational Methods for Solving the Electronic Schrödinger Equation

6.3.1 Electronic Structure Calculations

With the Born-Oppenheimer approximation, the Schrödinger equation can be simplified to deal with the electronic wavefunction equation that parametrically depends on the coordinates of the nuclei. Nevertheless, for various molecular systems solving the Schrödinger equation still faces the challenges arising from the many-body equation. Therefore, further approximations must be made in order to have the equation solvable. Historically, many different methods have been developed to address these challenges. With the rapid development of computational hardware and software over the past few decades, QM-based calculations can now be routinely performed using comprehensively programmed codes.

After understanding the basic concepts and theories, a researcher should be able to conduct molecular modeling jobs without the need of deep knowledge of the theories of QM and computer programming. Nevertheless, it is strongly advised that the use of any computational method implemented in a software package be considered based on sufficient background studies or consultation with experts. Treatment of a computational program as a black box is a dangerous practice and may easily result in false outcomes.

QM calculations provide detailed information about the electronic properties (electronic density, energy, and other types of distributions) of a specific molecular system. In solving the electronic Schrödinger equation, various approximation methods can be used and many of them are within the framework of *ab initio* quantum chemistry. Herein, **ab initio** means "from the beginning" or "from the first principles," since calculations only needs the positions of the nuclei and the number of electrons as inputs. General classes of *ab initio* methods are listed in Table 6.1. The details of each method can be found from specialized textbooks and scientific articles. It is worth noting that these methods are associated with different computational costs and chemical accuracies. While highly expensive methods can yield good accuracies, they are generally not advisable for modeling relatively large molecular systems. Conversely, when opting for less expensive low-level methods, the user needs to be aware of their limitations and inaccuracies. The selection of a particular *ab initio* method must strike a balance between computational costs and potential inaccuracies.

6.3.2 Hartree-Fock Methods

The Hartree-Fock (HF) methods are based on the Hartree-Fock scheme, which omits the instantaneous Coulombic electron–electron repulsion and only considers

Table 6.1 Classification of various *ab initio* methods.

Hartree-Fock	• Hartree-Fock (HF) • Restricted Hartree-Fock (RHF) • Unrestricted Hartree-Fock (UHF)
Post-Hartree-Fock	• Full configuration interaction (FCI) • Configuration interaction (CI) • Coupled cluster (CC) • Many-body perturbation theory (MBPT)
Multi-reference	• Multi-configurational self-consistent field (MCSCF, CASSCF, RASSCF, etc.) • Multi-reference configuration interaction (MRCI) • n-electron valence state perturbation theory (NEVPT) • Complete active space perturbation theory (CASPT) • State universal multi-reference coupled-cluster theory (SUMR–CC)

the average effect of electron–electron repulsion. This approximation is also known as the self-consistent field (SCF) method, as it iteratively treats each of the electrons in a molecule to make the calculations very effective without incurring high costs. However, the HF approach has several shortcomings. For example, the HF method cannot reasonably describe the breaking of a chemical bond and poorly predicts non-covalent interactions and transition states. Moreover, the HF results are strongly dependent on the basis sets used and sometimes tend to predict shorter bond lengths and overestimate energies. The intrinsic limitation of the HF method comes from the negligecting of electron correlation; therefore, it cannot be applied to the processes where the total number of paired electrons are changed. The lack of electron correlation can be addressed by an approach called full configuration interaction (FCI). The FCI scheme at the complete basis set (CBS) can give the exact, nonrelativistic Born-Oppenheimer electronic energy, but this method is extremely expensive and only applicable to small molecules. To solve the problem more efficiently, further approximations have been made, leading to the development of the post-Hartree-Fock methods. Among them, configuration interaction (CI) theory, coupled-cluster (CC) theory, and the many-body perturbation theory (MBPT) are well known. When the amount of dynamic correlation is small, it can be approximated by the perturbative methods. However, when the distinction between static and dynamic correlation becomes increasingly blurred, the wave function shows strong multireference character. These cases can be tackled by employing the multi-reference methods, which are computationally costly and only used when it is necessary. It is also worth knowing that expensive post-Hartree-Fock methods can be extended to a large system by dividing the system into two or more subsystems. Each subsystem is then treated at a different level of theory; a higher level of theory is applied to the subsystem of interest, while the environment is modeled at a lower level of theory such as molecular mechanics (MM). This approach is known as the quantum mechanics/molecular mechanics (QM/MM) method, which has been successfully used in the simulation of nanoscale systems such as drug-protein ensembles, polymers, and nanocarbon species.

6.3.3 Density Functional Theory Methods

The wavefunction-based post-Hartree-Fock methods can solve the Schrödinger equation for a many-body system with very good accuracies, but they face the challenges of poor scaling behavior as the size of the simulated system increases and the results are strongly dependent on basis set. Usually, post-Hartree-Fock methods scale with the fourth or higher power of the number of electrons in the calculated system, limiting their application only to small molecules. Starting from the 1980s, a method of completely different origin, called the **density functional theory** (DFT) approach, has emerged and achieved great success in electronic structure calculations on molecules and condensed matter. In 1998, a pioneering physicist in this field, Walter Kohn, was awarded the Nobel Prize in Chemistry with another renowned figure in theoretical chemistry, John A. Pople, for their contributions to the development of computational methods in quantum chemistry. Theoretically, the DFT method is rooted in the Thomas-Fermi theory, which was independently proposed by Llewellyn H. Thomas and Enrico Fermi around 1926. The Thomas-Fermi model provides a simple *ab initio* solution for calculating charge density and electron density in an atom or a molecule, but the results are not accurate because of several issues such as the expression of the kinetic energy being approximated. From a conceptual perspective, the Thomas-Fermi model gave birth to the field of DFT. In 1964, Hohenberg and Kohn formulated two theorems, which laid the foundation of modern DFT methods. The Hohenberg-Kohn (HK) theorems postulate that the ground-state electronic energy of an atom or molecule can be expressed exactly as a functional of its electron density (ρ). It is worth remarking that the term **functional** means a function of a function. The DFT approach can be expressed by a simple equation, $E_0 = E[\rho]$, meaning that the ground state-electronic energy, E_0, is a functional of the electron density ρ, where ρ is a function of the coordinates of the electrons in a system. In DFT theory, the electron correlation issue is solved in a much simpler way than those used in the post-Hartree-Fock methods. Therefore, DFT offers a cheaper but still accurate alternative to the post-Hartree-Fock methods. Nonetheless, the DFT approach is not perfect and fails completely in some situations. In DFT calculations, the total electronic energy, E, can be expressed as

$$E = T + V_{nuc} + V_{rep} + E_{XC} \qquad (6.15)$$

where T is the electronic kinetic energy, V_{nuc} is the nucleus-electron attraction, V_{rep} is the Coulomb repulsion among electrons, and E_{XC} is the exchange-correlation energy. In this equation, both the second and third terms, V_{nuc} and V_{rep}, are functionals of ρ. The first term T can also be expressed in terms of ρ, but the general expression is complicated and not completely known. The major problem with DFT is that the exact expression of the last functional, E_{XC}, in Eq. 6.15 is generally unknown. To deal with this deficiency, approximate functionals of E_{XC} are formulated to describe the exchange and correlation energies. In the development of practically useful DFT methods, approximations made on the exchange-correlation functional have resulted in hundreds of different density functionals. The large group of functionals can be classified by the type of approximation and ingredients included. As such, DFT methods exhibit a hierarchy called the **Jacob's ladder of DFT** [4]. This notion was first proposed by John P. Perdew in 2001. In Perdew's picture (see Figure 6.4), Hartree-Fock theory represents the earth. Each rung of the ladder contains new physical content that is missing from the lower rungs. Improved accuracy is attained as one climbs up toward the higher levels.

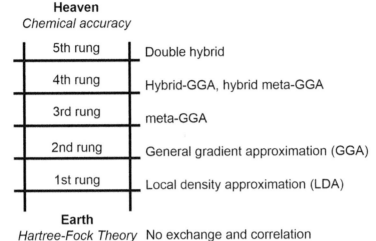

Figure 6.4 A metaphorical Jacob's ladder proposed by Perdew, illustrating different levels of accuracy for DFT methods.

The detailed methods at each rung are rather complex. Interested readers are suggested to read specialized monographs and review articles to gain deeper insights into the recent advancement in DFT methods [5–7].

Overall, the DFT methods have been extensively employed in modeling a wide range of molecular systems and condensed matter for decades. In the field of organic chemistry, DFT calculations are very useful; however, as the size of the modeled system increases, the computational demand can be intensive. Also, the use of different exchange functionals may result in different errors. The choice of an appropriate DFT method is therefore critical to the modeling studies. It is common that different research communities may have their favored methods (i.e., functionals). For example, an exchange correlation functional, namely B3LYP, has been extremely popular in the field of organic chemistry for several decades since it was first developed in the late 1980s. The adoption of other methods such as PBE, PBE1PBE, BP86, PW96, M06-2X, TPSSh, and ωB97X-D can also be frequently seen in the chemical literature. Most of these methods are available in modern quantum chemistry software packages.

6.3.4 Semi-Empirical Methods

Although *ab initio* and DFT methods are valuable modeling tools for understanding the electronic structures and energies of molecules, they are impractical in modeling very large systems such as biomolecules and nanosized molecular clusters. **Semi-empirical** methods can be used to simulate much larger systems at the quantum mechanical level. The development of semi-empirical methods has a long history, and these methods rely on the use of adjustable parameters in the Hamiltonian to solve the Schrödinger equation. Semi-empirical methods are typically three to four orders of magnitude faster than DFT and *ab initio* calculations, but still can achieve reasonable chemical accuracy through the tunning of the empirical parameters. Before the popular application of computer technology, some simple semi-empirical methods such as the Hückel model, extended Hückel method, and Pariser-Parr-Pople (PPP) method had been established mainly for modeling the π-electronic structures of conjugated molecules. Later, a method called neglect differential overlap (NDO) was developed as an extension of PPP to treat both σ and π electrons. There are two types of NDO methods, complete neglect of differential overlap (CNDO) and intermediate neglect of differential overlap (INDO). Based on them, some popular methods were further developed, such as MNDO, AM1, and PM3, which can predict the properties of various molecular systems within meaningful tolerance. In addition, a method based on approximation of DFT has been developed, known as the **density functional tight-binding** (DFTB) method. Most recently, an extended tight-binding method, GFN2-xTB, was developed by Grimme and co-workers. This method has demonstrated good accuracy, performance, and computational cost. With these improved semi-empirical methods, the modeling of large-scale systems, the sampling of extensive conformational space, and the dynamic simulations on relative a long timescale can be achieved at the QM level.

6.3.5 Basis Sets

One approximation applied in all *ab initio*, DFT, and semi-empirical methods is the use of basis sets to solve the molecular orbitals, since direct calculation of the mathematical form of a molecular orbital, $\psi(\mathbf{r})$, is impossible. To address the problem, a molecular orbital is expressed as a linear combination of a set of analytic functions, called **basis functions**, which are developed and optimized for individual atoms and can be viewed as representing the atomic orbitals, $\phi(\mathbf{r})$, of the atoms constituting the molecule.

$$\psi_i(\mathbf{r}) = \sum_{\alpha=1}^{n} \phi_\alpha(\mathbf{r}) C_\alpha \qquad (6.16)$$

As per Eq. 6.16, the task of solving a molecular orbital can be simplified as determining the expansion coefficient, C_α, of each basis function. Ideally, a complete basis set represents any kind of molecular orbital, but such a basis set has an infinite number of functions and practically not useful in reality. Over the past decades, many standard basis sets have been optimized and tested. In principle, the larger the basis set used, the more accurate the molecular orbitals solved. In practice, the computational expense grows rapidly with the size of the basis set being used. Therefore, a compromise must be sought between accuracy and cost.

Basis sets have been commonly constructed from the **Slater-type orbitals** (STOs) and **Gaussian-type orbitals** (GTOs). The STO basis sets were initially employed because they have the correct mathematical behavior. However, it takes a longer time to compute integrals when they are used. To solve the problem, a linear combination of GTOs can be used to mimic an STO, which is called an STO-nG basis set. Herein, n denotes the number of GTOs used. In QM calculations, the GTOs have been more widely used. Many modern QM software packages (e.g., *Gaussian, ORCA, GAMESS, Q–CHEM*) offer a wide range of pre-defined basis sets, which are classified by the number and types of basis functions they contain. These basis functions are composed of a linear combination of Gaussian-type functions, referred to as the **contracted functions**, and the component Gaussian functions are called **primitives**. Users may define their own basis sets or acquire basis sets from web-based platforms such as the Basis Set Exchange (https://www.basissetexchange.org) for special calculations.

There is a common nomenclature for various STO and GTO basis sets. The details can be found from standard textbooks on quantum chemistry and review articles. Minimal basis sets contain the minimum number of basis functions needed for each atom. For example, STO-3G is a minimal basis set that uses three gaussian primitives per basis function, which accounts for the "3G" in its name. Minimal basis sets are used for either qualitative analyses or calculating very large molecules. Basis sets can be made larger by increasing the number of basis functions per atom. For example, in the class of split valence basis sets (also called the Pople basis sets), 3-21G is a double-zeta split valence basis set, in which the numbers denote 3 GTOs for the inner shell (core), 2 GTOs for the inner valence, 1 GTO for the outer valence shell. 6–311G is a triple-zeta basis set, containing 6 GTOs for the core orbitals, 3 GTOs for the inner valence shell, and 2 different GTOs for the outer valence shell. The Pople basis sets can be modified to allow the orbitals to change size but not shape. This is done through the polarized basis set. In nomenclature, (d) or * is added after the name of the basis set to indicate polarization. For example, 6–31G(d) or 6-31G* is a basis set commonly used for calculating medium-sized molecular systems, and it is a 6-31G basis set with d functions added to the heavy (second period) atoms. The 6-31G(d,p) or 6-31G** basis set is another popular polarized Pople basis set that adds p functions to hydrogen atoms and d functions to heavy atoms. The Pople basis sets can also be modified by adding diffuse functions, allowing orbitals to occupy a larger region of space. This modification is useful when dealing with anions, excited states, molecules with lone pairs, and hydrogen bonded systems. The addition of diffuse functions can be indicated in the basis set name with + or ++ symbol(s) before G, for example, 6-31+G(d). Diffuse functions on hydrogen, however, seldom makes a significant difference in accuracy. In addition to the Pople type basis sets, other types of basis sets have been developed and frequently used in QM calculations (see examples in Table 6.2).

Table 6.2 Examples of different types of basis sets with varied sizes.

Pople type	Jensen type	Dunning type	Karlsruhe type
STO-3G	pc-1	cc-pVDZ	def2-SVP
3-21G	pc-2	cc-pVTZ	def2-SVPD
4-31G	pc-3	cc-pVQZ	def2-TZVP
6-31G(d,p)	pc-4	cc-pV5Z	def2-TZVPD
6-31+G(d,p)	aug-pc-1	aug-cc-pVDZ	def2-QZVP
6–311++G(3df,3pd)	aug-pc-2	cc-pCVDZ	def2-QZVPD

Basis sets can also be constructed from plane wave (PW) functions, which have been popularly used in computing periodic systems due to their high efficiency. However, PW basis sets are known to be inconvenient for molecular calculations. Delta functions provide another approach for constructing basis sets. The codes incorporating delta functions are simple, but thousands of functions are required to achieve accurate results, even for small molecules. In modeling heavy-atom systems, special basis sets such as pseudopotential or effective core potential basis sets are used to exclude the inactive atomic core electrons from an explicit treatment in QM calculations.

6.4 Applications of Electronic Structure Calculations

6.4.1 Computational and Visualization Software

In modern chemical research, computational modeling has achieved remarkable success in complementing and sometimes competing with experimental works. Hardware technologies such as supercomputers, parallel computing, and GPU acceleration have made complex QM calculations more and more efficient and affordable in these days. In the meantime, the optimization and improvement of various QM software packages have empowered general users to reliably carry out QM calculations at various levels of theory without comprehending in-depth knowledge about the codes and algorithms. This is analogous to that a person learning to safely drive a car by taking a driving course. How exactly the mechanical and electrical parts of the car work is not essential to the driving practice. Nevertheless, for researchers who want to set foot in the field of QM modeling and simulations, an in-depth understanding of fundamental quantum chemistry theories and sufficient computer skills are essential. There is no clearly defined threshold knowledge for practicing computational chemistry; however, the more knowledge one can gather the better the outcomes are. The following sections are intended to provide a brief introduction to basic QM computational skills and applications. To the beginners, the learning curve for these activities is not very steep, but caution should always be taken.

The first step of implementing a modeling task is to choose a suitable QM program for undertaking the computational work. Currently, there are many quantum chemistry software packages available. Some of them are free to either all users or academic researchers, while others require license fees to be paid. Before a particular QM program is selected, the user needs to carefully evaluate and compare the features, capacities, and limitations of different QM programs. Such information can be acquired by reading software manuals or surveying literature reports. Table 6.3 provides a selection of commonly used QM software packages.

Once a modeling task (research goal) is set up, the computational tasks can be conducted following the general workflow outlined in Figure 6.5. The first step of the workflow is to generate the initial nuclear coordinates of the molecular system to be studied. As discussed in the previous section, the QM calculations are based on the first principles (*ab initio*). Therefore, only the information about the positions of nuclei and the number of electrons is required for the calculations. For molecules

Table 6.3 Features and performances of selected software packages for QM calculations.

Software	License	Basis set	Semi-empirical	Hartree-Fock	Post-Hartree-Fock	DFT
ABINIT	Free	PW	No	No	No	Yes
ADF	Commercial	STO	Yes	Yes	No	Yes
CASTEP	Free (academic)	PW	No	Yes	No	Yes
CP2K	Free	STO, PW	Yes	Yes	Yes	Yes
Dalton	Free (academic)	GTO	No	Yes	Yes	Yes
Gamess(US)	Free	GTO	Yes	Yes	Yes	Yes
Gaussian	Commercial	GTO	Yes	Yes	Yes	Yes
MOLCAS	Free (academic)	GTO	Yes	Yes	Yes	Yes
MOLPRO	Commercial	GTO	No	Yes	Yes	Yes
MOPAC	Free	GTO	Yes	No	No	No
MPQC	Free	GTO	Yes	Yes	Yes	Yes
NWChem	Free	GTO, PW	No	Yes	Yes	Yes
ORCA	Free	GTO	Yes	Yes	Yes	Yes
PSI4	Free	GTO	No	Yes	Yes	Yes
Q-Chem	Commercial	GTO	Yes	Yes	Yes	Yes
Spartan	Commercial	GTO	Yes	Yes	Yes	Yes
TetraChem	Commercial	GTO	No	Yes	Yes	Yes
Turbomole	Commercial	GTO	No	Yes	Yes	Yes

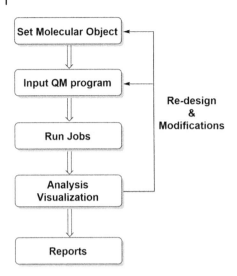

Figure 6.5 General workflow for executing a QM computational project.

with closed-shell electronic configurations, the calculations can be adequately dealt with by using the restricted method. If the electronic configuration of the molecular system is not a closed-shell one, the unrestricted method should be used. Also, the charge number and spin multiplicity of a molecular system need to be correctly determined. The charge number of a molecule is specified by a positive or negative integer that equals the total charge on it. For example, +1 indicates a singly charged cation, and −1 designates a singly charged anion, and 0 represents a neutral molecule. The spin multiplicity of a molecule is given by the formula $(2S + 1)$, where S is the total spin of the molecule. Note that electrons in a molecule take two spin states, α and β. To the total spin S, each electron contributes $+1/2$ (α spin) or $-1/2$ (β spin). In a system where the numbers of α and β electrons are equal (e.g., closed-shell), the total spin S is 0. Therefore, the spin multiplicity of such a system is 1, called a singlet. For a system, where there is an unpaired electron (e.g., a free radical), the total spin S is $+1/2$ and the spin multiplicity is 2 (doublet). A system containing two unpaired electrons with parallel spins possesses a total spin of $S = 1$ and a spin multiplicity of 3 (triplet), and so on.

For a molecular system, it is not easy to manually set up the coordinates of its nuclei unless its structure is very simple. Usually, a molecular model can be efficiently constructed through certain molecular editor software with a graphical user interface (GUI). Interested readers may want to check some popular molecular building software, such as *Avogadro*, *Chem3D*, *HyperChem*, *GaussView*, and *Spartan*. It is also worth mentioning that there are numerous web portals available to acquire experimentally determined or QM computed structures for a wide range of known organic molecules, for example, the Cambridge Crystallographic Data Centre (CCDC), PubChem, ChEMBL, ZINC20, and QM9. With the increasing application of AI technologies in chemical and biological sciences, such platforms are undergoing rapid growth.

A molecular geometry can be defined using different formats. For QM calculations, Cartesian x,y,z coordinates of nuclei are commonly used. Typically, the information of a molecular structure is stored in a text file that contains many lines. Each of the line begins with an atomic symbol or an atomic number, followed by the x, y, and z coordinates of this atom. The numerical values of x, y, and z

Cartesian coordinates of benzene

C	0.00000	1.40272	0.00000
H	0.00000	2.49029	0.00000
C	-1.21479	0.70136	0.00000
H	-2.15666	1.24515	0.00000
C	-1.21479	-0.70136	0.00000
H	-2.15666	-1.24515	0.00000
C	0.00000	-1.40272	0.00000
H	0.00000	-2.49029	0.00000
C	1.21479	-0.70136	0.00000
H	2.15666	-1.24515	0.00000
C	1.21479	0.70136	0.00000
H	2.15666	1.24515	0.00000

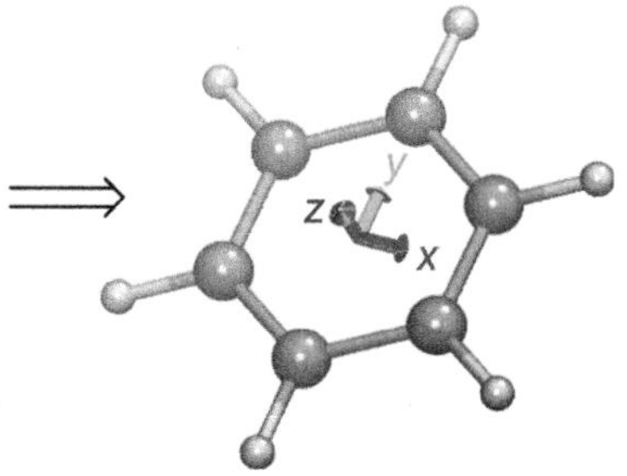

Figure 6.6 Cartesian coordinates of benzene and its molecular structure rendered by an open-source visualization program, *VMD*. The origin and the directions of x, y, and z axes are shown.

are in the unit of Angstroms (Å), defining the position of the atom in reference to an arbitrary origin. An atomic symbol and its coordinates are separated by spaces as illustrated by the Cartesian coordinates of benzene in Figure 6.6. In QM calculations, there is no need to define the bond connectivity among the atoms within a molecule, although in some cases bonds can be specified for certain purposes such as constraint analyses.

Once a set of nuclear coordinates is obtained, they can be processed by molecular visualization software to check that it gives the correct molecular structure. Some visualization programs can directly read Cartesian coordinates as inputs, while others need the Cartesian coordinates to be converted into readable formats. One commonly used translator program is called *Open Babel* (https://openbabel.org), which can read, write, and convert nearly all chemical file formats. As illustrated in Figure 6.6, a popular visualization software package, namely *VMD*, can clearly display the atoms and chemical bonds in a molecule according to pre-defined parameters. Sometimes, users can add and delete bonds through the GUI to better illustrate the molecular structure. A list of open-source molecular visualization software packages is provided in Table 6.4, some of which are used in generating the molecular graphics in this book.

6.4.2 Geometry Optimization

Various molecular properties, including energies, dipole moments, atomic charges, and orbital distributions, can be obtained from solving the Schrödinger equation of a single molecular geometry. The calculation based on a fixed molecular geometry is known as the **single-point calculation**. The lowest-energy solution for such a Schrödinger equation is called a **single-point energy**. On a PES,

Table 6.4 List of commonly used open-source molecular visualization software.

Software	Descriptions
Avogadro	An advanced molecular editor and visualizer designed for cross-platform (Windows, Linux, and MacOS) use in computational chemistry, molecular modeling, bioinformatics, materials science, and related areas. It has a user-friendly GUI to generate input files for various QM programs and can read many formats of output files to provide high-quality rendering of molecular properties such as structures, orbitals, and animation of vibrational modes. (https://avogadro.cc)
VMD	*VMD* stands for Visual Molecular Dynamics. It is a molecular visualization program for displaying, animating, and analyzing molecular structures and related properties (e.g., electron density maps, molecular orbitals, and trajectories) using 3D graphics. It is cross-platform running on Windows, Linux, and MacOS and supports built-in scripting. (https://www.ks.uiuc.edu/Research/vmd)
UCSF Chimera	A program for the interactive visualization and analysis of molecular structures and related data, including electron density maps, trajectories, and sequence alignments. It runs on Windows, Linux, and MacOS. (https://www.cgl.ucsf.edu/chimera)
Jmol	A Java-based program for displaying molecular, biomolecular, and crystallographic structures. It can generate surfaces, orbitals, and animations for molecular vibrations and reactions, and supports many chemical file formats, including the output files of many popular QM programs. (https://jmol.sourceforge.net)
CYLView	A visualization program that is very user-friendly to non-theoretical experimental chemists who wish to use computational chemistry as a practical tool. It displays molecular structures in various styles that are particularly suited for publication purposes. It runs on Windows and MacOS. (https://www.cylview.org)
QuteMol	An interactive, user-friendly molecular visualization program for displaying molecular and biomolecular structures with high-resolution and publication quality rendering. It can automatically create animated gifs of rotating molecules and read standard PDB files as inputs. The program runs on Windows and MacOS. (https://qutemol.sourceforge.net)
Gabedit	A GUI to many QM packages, such as *Gaussian, Gamess (US), MOLCAS, Molpro, MPQC, MOPAC, ORCA*, and *Q-Chem*. It can display a variety of calculation results including support for most major molecular file formats. It also has a molecule builder function for sketching molecular structures and examining them in 3D. It is cross-platform, running on Windows, MacOS, and Linux. (https://gabedit.sourceforge.net)

however, there typically exist a multitude of energy minima. The most stable one of these minima is called the **global energy minimum** and the others are referred to as **local energy minima**. In QM calculations, energy minimization algorithms can be applied to have an initial molecular geometry optimized; in other words, to find a minimum-energy structure(s). The computational task to find the exact nuclear coordinates for each of the minima is called **geometry optimization**. The outcome of a geometry optimization calculation is dependent on the initially guessed molecular geometry. If the guessed structure is close to the global energy minimum, the optimization can lead to the finding of it. However, if the initial geometry is far from the global minimum, the optimization steps may stop at a local energy minimum. In the latter case, the search for the global energy minimum requires either modifications of the initial geometry or implementation of suitable algorithms (e.g., meta-dynamics) to make the optimization steps cover a wide range of configurational and conformational space. Figure 6.7 depicts a simple 2D energy plot that illustrates the concept of local and global energy minima on a PES. Mathematically, an energy minimum on a PES must satisfy two criteria: (i) its first derivative is zero ($\partial E/\partial \mathbf{r} = 0$), meaning that it is a critical point of the function of energy and nuclear coordinates $\mathbf{r}$; (ii) its second derivatives are positive along all geometric parameters. To simply describe this, one can view an energy minimum as being located at the bottom of a concave up region on the 2D plot. The transition state (TS) is also a critical point ($\partial E/\partial \mathbf{r} = 0$), but its second derivative along one direction is negative; in other word, it is a maximum (concave down) along this direction.

Case study 1: Geometry optimization of cyclo[18]carbon
In Chapter 4, the synthesis and characterization of an intriguing carbon nanoallotrope, cyclo[18]carbon (C_{18}), is discussed (see Figure 4.45). Herein, a QM analysis of this nanocarbon species is demonstrated. First, let's perform a geometry

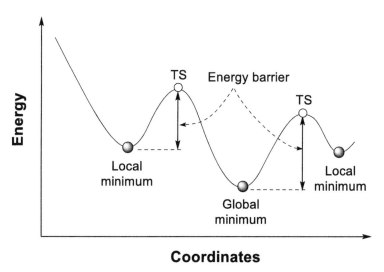

Figure 6.7 Illustration of energy minima and transition states on a 2D PES.

optimization of C_{18} using the QM methods implemented in the *ORCA* program (ver. 5.0.4) [8], which is free to download from https://orcaforum.kofo.mpg.de/ app.php/portal. As mentioned before, the initial geometry of a molecule can be constructed by using a suitable molecular structure editor. In this case study, the nuclear coordinates of C_{18} are constructed by the software, *Avogadro*, which can rapidly generate a preliminarily optimized geometry of C_{18} with a molecular mechanics (MM) approach (see Figure 6.8). From the *Avogadro* generated structure, Cartesian coordinates of the eighteen carbon atoms of C_{18} are obtained. It is worth mentioning that *Avogadro* has the GUI for generating input files for various popular QM programs, including *ABINIT, Gaussian, Molpro, ORCA, PSI4,* and *Q-Chem*. However, a plain text editor can also be used to create an input file for a QM program, which is recommended since users can directly and unambiguously specify input parameters and controls in this way.

An input file for QM calculations may take different formats and syntax, depending on the QM programs adopted. Users should carefully read the software manual to know how to create an input file for a specific computational task. For the software *ORCA*, the first line of an input file usually begins with "**!**," followed by keywords to specify the QM method, basis set, and job type. As exemplified in Figure 6.9, the first line of the input file uses the keyword "**HF**" (meaning Hartree-Fock) to specify the method of calculation, the keyword "**6–311G(d,p)**" to specify the basis set for the calculation, and the keywords "**Opt**" and "**Freq**" to indicate that the computational job begins with geometry optimization and then calculation of the vibrational frequencies of the optimized structure. The frequency calculation

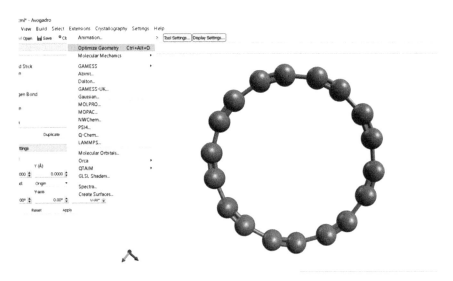

Figure 6.8 Screenshot of building a C_{18} molecule using the drawing tool and the geometry optimization function in *Avogadro* (ver. 1.2.0).

```
! HF 6-311G(d,p) Opt freq
%pal
 nprocs 16
end
%maxcore 4000
* xyz 0 1
C          0.566747      -0.975108       7.159429
C         -0.287349      -1.584510       6.576109
C         -1.481963      -2.227792       6.330923
C         -2.582648      -2.671009       6.513037
C         -3.763911      -2.984431       7.151325
C         -4.596162      -3.054076       8.013660
C         -5.211350      -2.890985       9.236761
C         -5.385747      -2.554471      10.375821
C         -5.147005      -1.991179      11.611432
C         -4.581947      -1.405964      12.494237
C         -3.600986      -0.706042      13.164202
C         -2.560868      -0.145954      13.377679
C         -1.296690       0.363097      13.168514
C         -0.268197       0.635985      12.612775
C          0.687675       0.715975      11.622350
C          1.223302       0.573977      10.557430
C          1.215738      -0.302966       8.173365
C          1.423604       0.187477       9.249177
*
```

Figure 6.9 An *ORCA* input file for optimizing the geometry of C_{18}, followed by frequency calculations at the HF/6–311G(d,p) level of theory.

can also be performed in a separate job after geometry optimization. Either way, they should be done at the same level of theory. If all the vibrational modes of the optimized geometry show positive frequencies, the obtained geometry is an energy minimum. The frequency calculation also provides properties related to thermal chemistry (e.g., zero-point energy, enthalpy, entropy, and free energy). In lines 2 to 4 of the exemplar input here, the number of CPUs and the memory assigned to each CPU for the calculation are specified. Users should define these hardware parameters according to the computer system they use. In the 6th line of the input file, a "*" symbol is placed at the beginning to indicate the section describing the nuclear coordinates of C_{18}. The keyword "**xyz**" specifies that nuclear positions are given as Cartesian coordinates (in the unit of Å). The numbers "**0 1**" denote the total charge of the molecule is zero (neutral) and the spin multiplicity is 1 (singlet). In the following lines, the Cartesian coordinates of the *Avogadro* generated C_{18} structure are indicated. The specification of nuclear coordinates is ended with another "*" symbol in the last line of the input file. Up to this point, a full input file for optimization and frequency calculations of C_{18} is created. This text file should then be saved with a filename extension ".inp" (e.g., C18-HF.inp) for QM calculations.

Running the input file can be executed by typing a command on a Linux or MacOS terminal or a Windows prompt, depending on the platform where the *ORCA* program is installed. Detailed installation instructions of *ORCA* can be found from https://www.orcasoftware.de/tutorials_orca. A simple command that runs an input file is "**orca C18-HF.inp > C18-HF.out**." Herein, "**C18-HF.out**" specifies the output file of the calculation. If the job is to be run on a supercomputer infrastructure, a job submission script (e.g., a shell script or a batch file) must be prepared. Understanding of the basic commands and script writing for Linux is therefore required before a user can skillfully handle submission of computational jobs.

After the calculation is finished, the *ORCA* program generates a series of files in the same directory as the input file. One of them has a filename extension of ".log" or ".out," which is the output file that stores the detailed information resulting from the QM calculations. This file can be directly read by a text editor or subjected to a visualization program such as *Avogadro* to plot the computed molecular structure as well as to perform various visualization analyses (e.g., orbitals, vibration modes). In addition to the .log file, there are other files generated for storing different types of information.

The C_{18} ring is a molecular carbon allotrope that has attracted broad theoretical and experimental interests since the early study by Hoffmann in 1966 [9]. One fundamentally important question about C_{18} is whether its molecular structure takes a polyynic or a cumulenic form (Figure 6.10). Various computational studies based on different theories have been conducted. According to the resonance theory, the C_{18} molecule can be treated as a resonance hybrid of the polyynic and cumulenic structures. The degree of π-electron delocalization can be quantified by an index called **bond length alternation** (BLA). For the C_{18} ring, the BLA is calculated as the average of the differences in length between the adjacent C–C type and C≡C type bonds. If the π-electrons of C_{18} are fully delocalized, all the

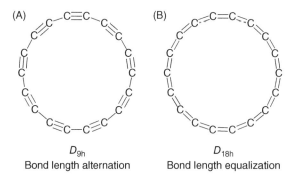

D_{9h}
Bond length alternation

D_{18h}
Bond length equalization

Figure 6.10 Two resonance structures for cyclo[18]carbon: (A) polyynic form and (B) cumulenic form.

bonds are equalized and the BLA index should be zero. In this case, C_{18} should favor the cumulenic structure. If the π-electrons of C_{18} are partially delocalized, the polyynic character would dominate the C_{18} structure, making the BLA index greater than zero. In 2019, Andersen and co-workers [10] performed experimental measurements using the atomic force microscopic (AFM) technique to confirm that the C_{18} ring favors a polyynic structure; that is, its cyclic carbon framework exhibits alternating short and long bonds.

Table 6.5 lists the bond lengths of two types of C–C bonds in C_{18} obtained from geometry optimization calculations at different levels of theories, including Hartree-Fock, post-Hartree-Fock, and DFT. All the computational jobs are completed by running input files like that shown in Figure 6.9, with suitable modifications on the keywords that describe the QM method and basis set. In reporting the results of QM calculations, the computational approach is typically presented as "method/basis set," for example, HF/6–31G(d) or B3LYP/def2-SVP.

In Table 6.5, most the DFT methods except ωB97X-D3 predict a cumulenic structure for the C_{18} ring with BLA = 0, which is inconsistent with the experimental results. The ωB97X-D3/def2-TZVP approach along with the Hartree-Fock and post-Hartree-Fock methods make a correct prediction of polyynic character; however, the detailed bond lengths and BLA values vary with the methods and basis sets used. Ågren et al. in 2019 performed systematic DFT and CASSCF calculations to confirm that the polyynic structure of C_{18} is the global minimum, whereas the cumulenic form is a transition state for the single–triple bond inversion process [11]. In 2020, Lu and co-workers [12] employed a very expensive QM method, CCSD/def-TVZP, to optimize the structure of C_{18}. Their results

Table 6.5 Comparison of the lengths of the two types of C–C bonds (d_1 and d_2) and BLA values of cyclo[18]carbon calculated at different levels of QM theories.

Level of theory	d_1 (Å)	d_2 (Å)	BLA (Å)
HF/6–31G(d)	1.379	1.198	0.181
HF/6–311G(d,p)	1.377	1.189	0.188
RI-MP2/cc-PVTZ[†]	1.324	1.250	0.074
B3LYP/6-31G(d)	1.285	1.285	0.000
BP86/def2-SVP	1.298	1.298	0.000
RI-R2BPLYP-D3BJ/def2-TZVP[‡]	1.281	1.281	0.000
ωB97X-D3/def2-TZVP	1.354	1.216	0.138

[†]The RIJCOX approximation is used along with cc-PVTZ/C and def2/J auxiliary basis sets.
[‡]The RI approximation is used along with def2-TZVP/C auxiliary basis set and D3BJ correction.

show that the two types of C–C bonds are 1.371 Å and 1.215 Å in length, and the BLA value is 0.156 Å. Therefore, the ωB97X-D3/def2-TZVP level of theory is an appropriate method for modeling the geometry of C_{18} in this case study. Obviously, the choice of a suitable computational method is highly important in QM studies. It should be performed based on in-depth understanding of the theories to be used. In the meantime, extensive comparison of different methods is always helpful. In certain cases of modeling studies, *ab initio* methods (e.g., Hartree-Fock) may predict outcomes better than many popular DFT methods; however, these outcomes need to be carefully considered. Beginners to this field must be aware that sometimes good outcomes may come from less robust theories.

Case study 2: Optimization of a Felkin-Ahn transition state
Transition-state properties are critically important to understanding organic reaction mechanisms. As mentioned in the beginning of this chapter, the proposal of a reasonable transition-state model allows the reactivity and selectivity of a type of organic reaction to be clearly understood and rationally predicted. In this case study, the QM calculations for finding reasonable transition states involved in stereoselective carbonyl addition reactions are demonstrated. Herein, let's use the addition of cyanide anion on (*S*)-2,3,3-trimethylbutanal as an example (Figure 6.11). This reaction can possibly yield two diastereomeric products. The major product can be predicted according to the Felkin-Ahn model previously described in Figure 6.1.

As discussed in the previous section, a transition state is a saddle point on a reaction trajectory. The calculations of the exact geometry and energy of a transition state can be achieved by executing certain transition-state finding algorithms implemented in QM software. Unlike ground-state geometry optimization, the successful search of a transition state critically depends on the initially guessed geometry. Two possible transition states can be empirically proposed for this addition reaction according to the Felkin-Ahn model. As illustrated in Figure 6.12A, a Felkin-Ahn transition state should allow the nucleophile (cyanide anion) to approach the carbonyl carbon of the addition substrate through a trajectory that encounters relatively small steric hindrance. With this model, a Felkin-Ahn product can be deduced and optimized initially. Herein, the structure of this product is optimized by a DFT method using the ωB97X-D3 functional in conjunction

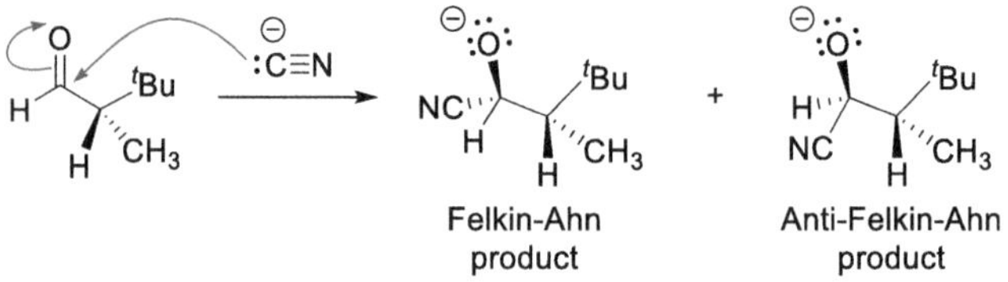

Felkin-Ahn
product

Anti-Felkin-Ahn
product

Figure 6.11 Addition reaction of cyanide anion on (*S*)-2,3,3-trimethylbutanal.

with the 6-31+G(d) basis set. Note that the "+" sign indicates the addition of diffuse functions to the basis set, which is used in this case in view of the anionic character of this reaction system.

Optimization of the ground-state geometry of the Felkin-Ahn product results in the structure shown in Figure 6.12, in which the carbonyl carbon and cyanide carbon form a C–C single bond with a bond length of 1.546 Å. With this structure in hand, the related Felkin-Ahn transition state is assumed to be located on the trajectory where this C–C bond is progressively elongated. Such an assumption is based on the Principle of Microscopic Reversibility, which states that both the forward and reverse paths of a reaction are identical and hence have the same transition state. As the geometry of the product can be more easily determined in this addition reaction, it is much easier to search the transition state along the reverse reaction direction.

There are numerous methods for locating approximated transition states. In this case study, a simple method called a relaxed PES scan is used. This type of calculation allows the PES to be explored by optimizing molecular geometries with varied parameters (e.g., bond lengths, bond angles, and dihedral angles). In the *ORCA* software program, a relaxed scan calculation can be carried out by using the keyword "**%geom scan**" to specify the parameters to be scanned. As shown in the input file in Figure 6.13, an inexpensive semi-empirical method (PM3) is adopted for a rapid relaxed scan along the C–C bond that is formed during the addition reaction. The scan begins with the Felkin-Ahn product and the length of the reacting C–C bond is progressively elongated from 1.55 Å to 2.65 Å. In the third line of the input file, "**B**" indicates that the scanned parameter is a bond length. "**0**" and "**22**" are the numbering of the two atoms in the Cartesian coordinates, which

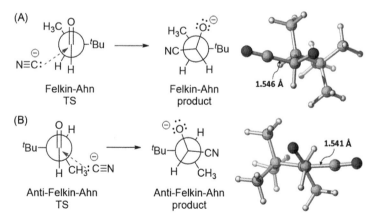

Figure 6.12 Two proposed transition states and their corresponding products for the addition of cyanide anion to (*S*)-2,3,3-trimethylbutanal. Product structures are optimized at the ωB97X-D3/6-31+G(d) level and rendered with the *CYLView* program.

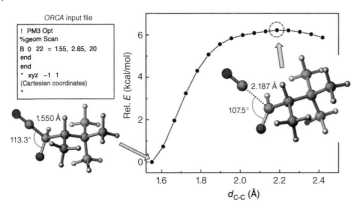

Figure 6.13 Relaxed PES scan for searching an approximated transition state. Calculations are done at the semi-empirical PM3 level of theory starting from a DFT optimized product structure.

define the bond to be scanned. One should be made aware that the program *ORCA* numbers the first atom in the Cartesian coordinate as "**0**" not "**1**." Other QM software packages may have different ways of numbering the atoms. Users should read the software manual carefully for such details. The following numbers "**1.55**" and "**2.65**" specify the range of bond length scan (in the unit of Å). Finally, the number "**20**" instructs the program to take twenty scan steps.

After the relaxed scan, the optimized geometry and energy resulting from each scan step can be found from the output files. In this case, a smooth energy profile is plotted based on the first 16 scan steps (see Figure 6.13). In this plot, the 12th step appears to be an energy maximum, and the position of the cyanide anion relative to the carbonyl carbon in this structure shows good agreement with the Bürgi-Dunitz trajectory. This structure is therefore deemed as a reasonable approximation of the Felkin-Ahn transition state. To compute the precise geometry and energy of the transition state, a transition-state optimization job is subsequently conducted with *ORCA*. Part of the input file is shown below.

```
! wB97X-D3 6-31+G(d) OptTS Freq
%pal
 nprocs 32
end
%maxcore 4000
* xyz -1 1
(Cartesian coordinates)
*
```

In the first line of the above input file, the keywords "**wB97X-D3**" and "**6-31+G(d)**" define the level of theory for the QM calculations. The keyword "**OptTS**" instructs the program to perform transition-state geometry optimization based on the Cartesian coordinates of a guessed transition-state structure. The following keyword "**Freq**" tells the program to calculate the vibrational frequencies of the optimized transition state structure. As discussed before, the frequency calculations are essential to validate an optimized saddle point as a true transition state (i.e., one imaginary frequency for the vibration along the reaction trajectory). It also provides relevant thermodynamic properties for the optimized transition state. Lines 2 to 5 specify the number of CPUs and the size of memory requested for the calculations, and line 6 defines the format of the atomic coordinates, total charge, and spin multiplicity.

After running the transition-state optimization and frequency calculations, the *ORCA* program generates an output file (.log file) with a section listing the vibrational frequencies of the optimized structure. As shown in Figure 6.14, there is only one imaginary mode with a frequency at -217.25 cm^{-1} for this optimized transition state. This result confirms that the optimized structure is a true transition state on the PES. It is worth mentioning that *ORCA* and some other QM programs present imaginary frequencies as negative values in the output files. Nevertheless, readers should be aware that mathematically such frequencies are imaginary numbers, not negative real numbers.

Besides the Felkin-Ahn transition, an anti-Felkin-Ahn transition state is conceivable following the rationalization depicted in Figure 6.12. The corresponding anti-Felkin-Ahn product is then deduced and modeled by QM calculations. With the optimized anti-Felkin-Ahn product, a relaxed PES scan is performed in a manner similar to the PES scan on the optimized Felkin-Ahn product. The energy profile of this PES scan is shown in Figure 6.15. Unlike the case of the Felkin-Ahn transition state, the total energies in Figure 6.15 exhibit a monotonously

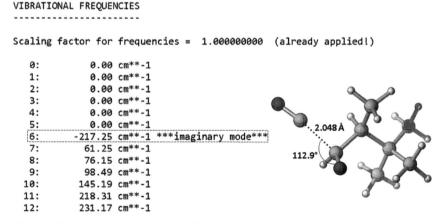

```
----------------------
VIBRATIONAL FREQUENCIES
----------------------

Scaling factor for frequencies =  1.000000000  (already applied!)

     0:         0.00 cm**-1
     1:         0.00 cm**-1
     2:         0.00 cm**-1
     3:         0.00 cm**-1
     4:         0.00 cm**-1
     5:         0.00 cm**-1
     6:      -217.25 cm**-1 ***imaginary mode***
     7:        61.25 cm**-1
     8:        76.15 cm**-1
     9:        98.49 cm**-1
    10:       145.19 cm**-1
    11:       218.31 cm**-1
    12:       231.17 cm**-1
```

Figure 6.14 Part of the *ORCA* output file showing the vibrational frequencies calculated for the optimized Felkin-Ahn transition state. The structure of the optimized transition state is shown on the right-hand side with the key geometric parameters highlighted.

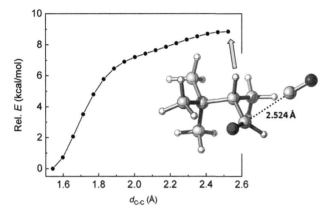

Figure 6.15 Relaxed PES scan of the anti-Felkin-Ahn product along the C–C bond between cyanide carbon and carbonyl carbon. Calculations are done at the semi-empirical PM3 level of theory.

increasing trend as the C–C bond length is elongated from 1.54 Å to 2.64 Å. There is no clear saddle point present on this PES, suggesting that a meaningful transition state along this reaction trajectory does not exist. The search of transition states leading to the formation of the anti-Felkin-Ahn product hence requires a more systematic conformational and configurational exploration. Overall, this case study exemplifies one of the many useful approaches for transition state searches and optimization. QM modeling of transition states not only allows reaction mechanisms (e.g., reactivity and selectivity) to be quantitatively understood, but offers theoretical guidance for the development of new reaction methodologies and functional molecules (e.g., drugs, receptors, and sensors). For complex organic reaction mechanisms, the comprehensive search of important transition states is not a trivial task. There are numerous computational methods and modeling "tricks" that can be found from the vast body of computational research articles and textbooks. Interested readers are directed to study them for better practice in this field.

6.4.3 QM Simulations of Other Molecular Properties

Case study 3: QM simulations of molecular magnetic properties

QM calculations can be used to simulate and interpret various molecular spectroscopic data, ranging from IR, Raman, NMR, UV-Vis absorption, and fluorescence. The calculations of IR and Raman spectroscopic properties can be achieved by adding the keyword "**Freq**" in an *ORCA* input file, which allows the molecular vibrational modes to be computed. The output file of a frequency calculation job can be visualized using *Avogadro* as a GUI to produce a simulated vibrational spectrum and to visualize specific vibrational modes through the "animation" function in the software. As mentioned before, frequency calculations also provide various thermodynamics properties. Magnetic properties such as NMR spectroscopic data can be simulated through QM calculations. In routine lab work, NMR analysis is an indispensable characterization tool for organic chemists to elucidate molecular structures, monitor chemical processes, and determine reaction mechanisms. Molecules with complex structures, however, tend to give NMR spectral patterns that cannot be easily deciphered. QM-based simulations of NMR properties can be very helpful for these studies.

For π-conjugated cyclic molecules, quantitative understanding of their aromaticity or anti-aromaticity is important. As introduced in Chapter 4, many exotic carbon allotropes show intriguing electronic and optical properties due to their unique aromatic character. Reliable metrics for evaluating aromaticity have been developed and tested for many decades. In Chapter 1, a computational method called the **nucleus independent chemical shift** (NICS) is briefly discussed as a magnetism-based criterion for assessing and quantifying aromaticity/anti-aromaticity. This method was first proposed by Paul v. R. Schleyer [13] in 1996 and has thereafter attracted enormous attention from the organic chemistry community, mainly owing to its user-friendliness, intuitive interpretation, and relatively low computational complexity.

In this case study, NMR calculations of benzene are demonstrated by running a single-point job through the *ORCA* program (see Figure 6.16). To let the *ORCA* program run NMR simulations, a keyword "**NMR**" needs to be placed at the beginning of the input file. To obtain NICS values, dummy atoms are used as

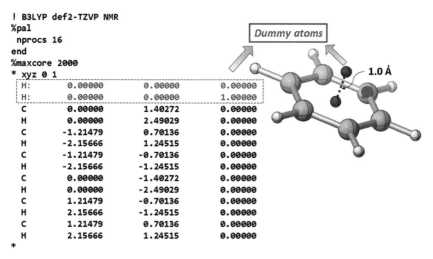

Figure 6.16 An exemplar *ORCA* input file for computing the NICS(0) and NICS(1.0) values of benzene. A pictorial illustration of the dummy atoms placed within and above the benzene ring is shown on the right-hand side.

magnetic probes and they are placed at suitable positions with respect to the ring of interest. The original proposal of NICS by Schleyer is to simply compute the absolute shielding (isotropic chemical shift) of a dummy atom placed at the non-weighted geometric center of the ring to be examined, and its negative value is defined as the NICS of the ring. This method nowadays is known as the NICS(0), since the dummy atom is located in the plane of the molecule with a vertical distance of zero. The NICS(0) is an easy and straightforward aromaticity criterion, but its application faces some problems that may mislead the interpretation of the outcomes. One of the problems is ascribed to σ-contamination, which arises from the convolution of σ- and π-electronic effects. To address this problem, a revised method called NICS(1.0) was devised later [14]. This method has the dummy atom placed vertically above the geometric center of the ring at a distance of 1.0 Å. In this way, the effects of σ-electrons can be significantly reduced. As a result, the calculated magnetic shielding can better describe the behavior of the π-electrons and hence the aromaticity of the system.

In the example shown in Figure 6.16, all the hydrogen and carbon atoms of the benzene ring are in the *xy*-plane of the Cartesian coordinates. This setting makes it easy to decide the positions of the dummy atoms for NICS(0) and NICS(1.0) calculations. If the ring structure is defined by more complex coordinates (e.g., non-planar), the location of a geometric center needs to be carefully determined. Some molecular modeling software packages (e.g., *GaussView* and *CYLView*) are equipped with the function of calculating geometric centers for rings and cages. In computational chemistry, dummy atoms refer to artificial atoms commonly used to represent points in space. Dummy atoms are useful in defining structures or constraints without placing actual atoms around the molecule to be modeled. Different QM software packages have different syntax for specifying dummy atoms in their input files. In *Gaussian*, a dummy atom is denoted as "**Bq**," while in the *ORCA* program a dummy atom can be written as "**H:**" in the section of nuclear coordinates.

After running the input file as shown in Figure 6.16, the *ORCA* program stores the calculated NMR properties in a file "**xxx**_property.txt" (**xxx** is the name of the input file). In the output file, the magnetic shielding tensors, tensor eigenvectors, and tensor eigenvalues for all the nuclei are listed in a section as illustrated in Figure 6.17. Users can find the absolute shielding value, P (iso), of a particular nucleus. Just a reminder that the first atom in the list is numbered as 0 not 1. In this computational job, the two dummy atoms yield isotropic chemical shifts at 8.0 ppm and 10.0 ppm, respectively. The NICS(0) and NICS(1.0) values of benzene calculated at the B3LYP/Def2-TZVP level are accordingly determined as −8.0 and −10.0. These values are consistent with the literature-reported NICS values for benzene. For example, Schleyer et al. reported the NICS(0) of benzene as −8.9 and NICS(1.0) as −10.6 at the IGLO-PW91/IGLO-III level of theory [15]. This case

```
Source density: 1 SCF
Nucleus: 0 H                          First dummy atom
Shielding tensor (ppm):
                  0          1          2
    0        4.139539  -0.000002   0.000013
    1        0.000007   4.136586  -0.000004
    2       -0.000014  -0.000008  15.846250
P Tensor eigenvectors:
                  0          1          2
    0        0.000786   1.000000  -0.000001
    1       -1.000000   0.000786  -0.000001
    2       -0.000001   0.000001   1.000000
P Eigenvalues:
                  0          1          2
    0        4.136586   4.139539  15.846250
P(iso)   8.040792                      NICS(0) = -8.0
Nucleus: 1 H                          Second dummy atom
Shielding tensor (ppm):
                  0          1          2
    0        0.313572   0.000002   0.000005
    1       -0.000001   0.312716  -0.000009
    2       -0.000012  -0.000008  29.506417
P Tensor eigenvectors:
                  0          1          2
    0        0.000361  -1.000000  -0.000000
    1       -1.000000  -0.000361  -0.000000
    2       -0.000000  -0.000000   1.000000
P Eigenvalues:
                  0          1          2
    0        0.312716   0.313572  29.506417
P(iso)   10.044235                     NICS(1.0) = -10.0
```

Figure 6.17 Part of the *ORCA* output file for the NICS calculations of benzene. The absolute isotropic chemical shifts of the two dummy atoms are highlighted.

study just demonstrates a simple approach to evaluate the aromaticity and anti-aromaticity of π-conjugated cyclic compounds. Nowadays, the NICS method has been considerably expanded to describe and quantify aromaticity and anti-aromaticity in a more accurate and insightful manner [16]. Without any doubt, the NICS method has become increasingly sophisticated and therefore needs a certain level of expertise to properly handle the computational tasks and outcomes. Beginners in this field are advised to establish sufficient background knowledge prior to modeling studies or to collaborate with experts to avoid misinterpreting the results of NICS calculations.

Case study 4: QM simulations of electronic absorption properties
Besides the ground-state properties, QM calculations may provide understanding of the excited-state properties. Many π-conjugated molecules show intensive absorption in the ultraviolet and visible (UV-Vis) regions of the spectrum, owing to facile vertical electronic transitions from the ground to excited electronic states. The properties of excited states result from light-matter interactions and hence require the time-dependent (TD) nature of the electromagnetic waves to be taken

into account in modeling studies. This means that the computation of excited-state properties must solve the time-dependent Schrödinger equation instead of the time-independent one. As such, the modeling and simulation of excited states are far more challenging than the ground state. Over the past few decades, time-dependent density functional theory (TD-DFT) [17] has emerged as a popular theoretical tool for modeling various excited-state properties. In this case study, the UV-Vis absorption spectrum of cyclo[18]carbon in the gas phase is simulated using a TD-DFT method. There is a plethora of density functionals available for TD-DFT calculations. As a rule of thumb, the hybrid functionals (e.g., B3LYP, PBE0, or M06) tend to provide accurate estimates for organic and inorganic systems. In the cases where charge transfer is significant, range-separated hybrids (e.g., CAM-B3LYP, ωB97X-D) are recommended. In TD-DFT calculations, the basis set should be relatively large in order to obtain meaningful results. For example, a triple zeta basis set is often considered as a pertinent choice. If charge-transfer states are involved, the inclusion of diffuse functions in the basis set is necessary.

An exemplar TD-DFT input file is illustrated in Figure 6.18. The first line specifies the use of a hybrid B3LYP functional in conjunction with the Def2-TZVP basis set for the calculations. In the second line, a keyword "**%TDDFT**" is used to let the *ORCA* program know to run excited-state calculations. The third line uses

```
!B3LYP DEF2-TZVP
%TDDFT
    NROOTS    50
END
%pal
  nprocs 16
end
%maxcore 4000
* xyz 0 1
  C    0.56182345093427    -0.96618080999199     7.17429725346134
  C   -0.30281615933748    -1.58612316820061     6.58447678967675
  C   -1.47334395099328    -2.22206188392547     6.34468601403337
  C   -2.58758936202219    -2.67385583674921     6.52978994688021
  C   -3.74691586732221    -2.98193571258917     7.15700286828055
  C   -4.59153217414718    -3.05051541702005     8.02982539043904
  C   -5.19801986043848    -2.88589982803569     9.22867006096293
  C   -5.37566281978861    -2.54155737719679    10.38191769148902
  C   -5.14143558008906    -1.98728950937483    11.59438125729690
  C   -4.56838703388540    -1.39585873172518    12.48975628227434
  C   -3.60361529900793    -0.71275751989700    13.14909546438660
  C   -2.54821069900572    -0.14813260608833    13.36626449901969
  C   -1.30631230848687     0.34977108792922    13.16124234021800
  C   -0.26414691289992     0.62656327778386    12.59798675502856
  C    0.67333839030125     0.70733599869790    11.62502249291920
  C    1.21511672133618     0.56635559401982    10.54505920996366
  C    1.19933552951252    -0.30582020724389     8.16908767229779
  C    1.41061693534010     0.18998664960742     9.25966401137199
*
```

Figure 6.18 An *ORCA* input file for running TD-DFT calculations of cyclo[18]carbon.

a keyword "**NROOTS**" to specify how many excited states to be calculated in this job. In this TD-DFT simulation, a total of 50 excited states of cyclo[18]carbon is computed based on the geometry previously optimized from the ωB97X-D3/Def2-TZVP level of theory.

After running the job, an output file is produced, which contains numerous sections listing the details of atomic charges, orbital charges/energies, excited-state properties, and so forth. An easy way to visualize the TD-DFT simulated UV-Vis absorption spectrum is to open the output file with *Avogadro* and then click the "Spectra" option under the "Extension" menu (see Figure 6.19).

More detailed electronic transitions resulting from the TD-DFT calculations can be found from a set of tabulated data in the output file as exemplified in Figure 6.20. In this section of data, the first column indicates the excited states, the second column denotes the vertical electronic transition energy in the unit of cm^{-1}. Note that the TD-DFT calculations here only consider singlet-to-singlet electronic transitions. The third column shows the wavelengths corresponding to these electronic transitions, which can be compared with experimentally measured UV-Vis absorption data if they are available. The fourth column provides the values called **oscillator strengths** (f_{osc}) for the electronic transitions between the ground and excited states. In spectroscopy, oscillator strength is a dimensionless quantity that measures the probability of absorption or emission of electromagnetic radiation. The cyclo[18]carbon structure possesses a high degree of symmetry (D_{9h}). As a result, many of the electronic transitions show oscillator strengths equal to zero, meaning that these transitions are symmetry forbidden and not likely to be experimentally

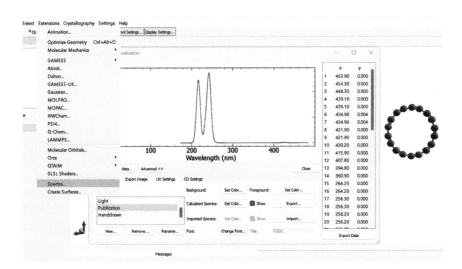

Figure 6.19 Screenshot of the TD-DFT simulated UV-Vis spectrum of cyclo[18]carbon visualized with *Avogadro*.

```
---------------------------------------------------------------------------
         ABSORPTION SPECTRUM VIA TRANSITION ELECTRIC DIPOLE MOMENTS
---------------------------------------------------------------------------
State   Energy    Wavelength  fosc        T2         TX        TY        TZ
        (cm-1)    (nm)                   (au**2)     (au)      (au)      (au)
---------------------------------------------------------------------------
  1    21557.7    463.9    0.000000000   0.00000   -0.00002   0.00003   0.00004
  2    22012.4    454.3    0.000000006   0.00000   -0.00004   0.00007   0.00030
  3    22306.2    448.3    0.000000000   0.00000   -0.00001   0.00002   0.00001
  4    22775.7    439.1    0.000000001   0.00000   -0.00009  -0.00005  -0.00002
  5    22776.1    439.1    0.000000001   0.00000    0.00003  -0.00004  -0.00010
  6    23536.6    424.9    0.003969057   0.05552    0.04481  -0.06098  -0.22313
  7    23536.9    424.9    0.003971951   0.05556   -0.21143  -0.10319  -0.01429
  8    23704.4    421.9    0.000000013   0.00000   -0.00001  -0.00020  -0.00038
  9    23704.5    421.9    0.000000020   0.00000    0.00049   0.00021   0.00003
 10    23795.6    420.2    0.000000078   0.00000   -0.00051   0.00084  -0.00034
 11    24043.6    415.9    0.000000012   0.00000   -0.00038  -0.00017   0.00001
 12    24521.5    407.8    0.000009704   0.00013    0.00454  -0.00983   0.00360
 13    25331.1    394.8    0.000000013   0.00000    0.00034   0.00022   0.00007
 14    27711.3    360.9    0.000000015   0.00000   -0.00010   0.00014   0.00038
 15    37845.4    264.2    0.000000008   0.00000    0.00006   0.00011   0.00024
```

Figure 6.20 Part of the output file for the TD-DFT calculations for cyclo[18]carbon, tabulating the details of electronic transitions between the ground and excited states.

observed. The strongest oscillator strengths correspond to absorption wavelengths at 243 nm and 216 nm, which match the region where typical $\pi \rightarrow \pi^*$ transition bands would appear. These results generally agree with those calculated by other TD-DFT methods [18]. In addition to the simulated UV-Vis spectral data, TD-DFT calculations also provide useful information about the orbital contributions to each of the electronic transitions. The electronic transition properties can be visualized by several ways (e.g., natural transition orbitals). Interested readers are suggested to read relevant literature for details.

6.5 Computational Methods Based on Empirical Force Field Models

6.5.1 Introduction to Molecular Mechanics

QM-based computational methods can provide sufficient chemical accuracy for modeling small- to medium-sized molecular structures, but they are incapable of studying large molecular and supramolecular systems. One advantage that makes QM methods accurate and reliable is that electrons are taken into consideration in the computational models. On the other hand, it is also because of this, the QM approach must face the challenge of solving the many-body Schrödinger equation, which is computationally expensive. The QM methods encounter limitations in dealing with molecular systems containing a significant number of atoms (e.g., polymers and proteins). To overcome this barrier, another class of modeling methods has been extensively used, which are known as the **force field** or **molecular mechanics** (MM) methods. The MM approach calculates the energy of a molecular system as a function of the nuclear positions only, whereas electrons are neglected in the modeling. In this way, the problem of solving the many-body Schrödinger equation is averted. With such an approximation, the MM methods are capable of modeling very large molecular and supramolecular systems with high efficiency and good accuracy. Some developed MM methods can even provide outcomes as accurate as high-level QM calculations without consuming significant computational times and resources. On the other hand, the neglect of electrons in the modeling makes the MM approach completely unable to describe electronic distribution.

Like the QM methods, the MM modeling works on the basis of several approximations. The first is also the Born-Oppenheimer approximation, which allows the energy to be calculated as a function of the nuclear coordinates. The second approximation is to treat molecular systems as simple models containing stretching bonds, bending bond angles, rotational motions about single bonds, and other non-bonded interactions (e.g., electrostatic and van der Waals forces). The third approximation that enables the MM modeling is called **transferability**. In calculating the energy of a molecular system, a set of functions and empirical parameters are generated, which is called the force field. A type of force field developed and tested on a relatively small number of molecular systems is believed to be applicable to a much wider range of molecular systems. In particular, parameters derived from small molecules can be applied to the modeling of much larger systems such as proteins and polymers.

6.5.2 Basics of Force Fields

There have been many types of force fields developed for MM modeling. Most of them take the general form of the following equation.

$$V\left(r^N\right) = \sum_{bonds} \frac{k_i}{2}(l_i - l_{i0})^2 + \sum_{angles} \frac{k_i}{2}(\theta_i - \theta_{i0})^2 + \sum_{torsions} \frac{V_n}{2}\left(1 + \cos\left(n\phi - \delta\right)\right)$$
$$+ \sum_{i=1}^{N}\sum_{j=i+1}^{N}\left(\frac{q_i q_j}{4\pi\varepsilon_0 r_{ij}} + 4\varepsilon_{ij}\left[\left(\frac{\sigma_{ij}}{r_{ij}}\right)^{12} - \left(\frac{\sigma_{ij}}{r_{ij}}\right)^{6}\right]\right) \qquad (6.15)$$

In Eq. 6.15, the term on the left-hand side, $V(\mathbf{r}^N)$, denotes the potential energy, which is a function of the nuclear coordinates ($\mathbf{r}^N$). On the right-hand side of this equation, there are four basic components that account for various intramolecular and intermolecular interactions. The first term describes the energy due to the interactions between pairs of bonded atoms, where l_i refers to the bond length of the ith pair of atoms and l_{i0} is the bond length at equilibrium. Mathematically, this term is equivalent to that for a harmonic oscillator model that obeys the Hooke's law. The second term describes the bending interactions in a three-atom

"A–B–C" system, and θ is the bond angle between them. The third term describes the contribution of bond rotation in a four-atom "A–B–C–D" system, in which φ is the dihedral (torsion) angle between them. The fourth term of Eq. 6.15 comprises two components. The first one is the Coulomb's law expression describing the electrostatic interactions between two non-bonded particles. The second is the van der Waals force (attraction and repulsion) described by the Lennard-Jones potential. The physical meanings of the intramolecular and intermolecular forces mentioned above are pictorially presented in Figure 6.21.

In MM modeling, a force field must be specified in terms of the functional forms and the parameters as shown in Eq. 6.15. A variety of force fields have been developed to reproduce structural properties as well as to predict other properties such as spectral data. It is worth noting that, unlike the first-principle methods, force fields are empirical and there are no "correct" forms for them. Different types of systems therefore require different force fields to simulate them. In general, there are three classes of force fields used in MM modeling studies. The Class I force fields have the bond stretching and angle bending described by simple harmonic motions. For example, AMBER, CHARMM, GROMOS, and OPLS. Class II force fields add anharmonic terms to the potential energy for bond stretching and angle bending, and they have cross-terms to describe the coupling between adjacent bonds, angles, and torsions. Examples of Class II force fields are MMFF94 and UFF. In the Class III force fields, special effects of organic chemistry such as polarization, stereoelectronic, and Jahn-Teller effects are explicitly added. Examples of Class III force fields are AMOEBA and DRUDE. The proper choice of a force field for MM modeling is dependent on the system to be simulated.

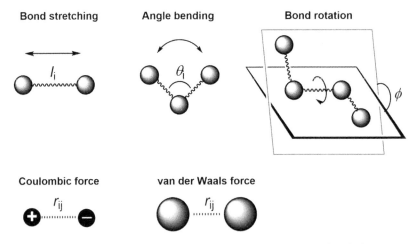

Figure 6.21 Pictorial illustration of the different bonded and non-bonded components in a force field equation.

A perfect force field does not exist. Most of the force fields used in MM modeling are designed to simulate molecular structures near equilibrium. They can model molecules and solvent–solute interactions, but are not sufficient to model chemical processes such as bond formation, bond breaking, and charge transfers. Each force field can provide accurate results for a certain range of chemical systems but becomes less accurate for others. The applicability of a force field relies on the data used for its parameterization. The robustness of a force field is also an important consideration in the MM modeling. A less robust force field may give very accurate simulation results for one type of system, but very poor for others. A robust force field may deliver good results for a wide range of molecular systems, but it is not always the most accurate among other force fields. Table 6.6 summarizes the force fields included in commonly used MM modeling software.

MM methods can be applied to many types of modeling studies. First, they can be readily used to compute the energies and optimized structures of various molecular systems, especially, relatively large biomolecules and polymers. Second,

Table 6.6 Summary of various force fields and their usage.

Force field	Primary use
AMBER	Proteins, nucleic acids
AMOEBA	Proteins
CHARMM	Proteins, nucleic acids
CHEAT	Carbohydrates
CPE	Molecular liquids
DREIDING	Organic and bioorganic molecules
ECEPP	Peptides
EFF	Hydrocarbons
ENZYMIX	Biological molecules
GROMACS	Molecular dynamics
GROMOS	Molecular dynamics
MM1, MM2, MM3, MM4, MM+	Organic molecules
MMFF	Organic molecules and biomolecules
MOMEC	Transition metal coordination systems
OPLS	Liquid simulations
PFF	Molecular dynamics
QCFF/PI	Conjugated molecules
ReaxFF	Dynamic simulations of reactions
SIBFA	Small molecules, proteins
SYBYL	Organic and bio-organic molecules
UFF	Full periodic table
YETI	Small molecule-protein complexes

incorporation of Newton's equations into the MM methods allows molecular motions (vibrations and movements) in a solvent to be simulated. Third, the MM methods can be applied to Monte Carlo simulations to statistically describe molecular systems of interest. Moreover, the MM methods can be used for molecular docking studies, where the orientations and conformations of a ligand molecule (inhibitor) bound to various active sites of a protein are calculated.

6.5.3 Basics of Molecular Dynamics Simulations

Molecular dynamics (MD) simulations are computational techniques for analyzing the motions and energies of atoms and molecules over a period of time. Its development was largely motivated by the desire to understand the structure–function relationships for macromolecules (e.g., proteins and nucleic acids) at the molecular level. Historically, the modeling of molecular dynamics started long before the invention of computers. In the mid-1800s, structural models were proposed for molecules to explain their chemical and physical properties. First attempts at modeling interacting molecules were made by van der Waals in his pioneering work, *About the Continuity of the Gas and Liquid State*, published in 1873. In this work, he used a simple equation that approximates molecules as interacting spheres to predict the transition of gas to liquid. Modern MD simulations still adopt many of the concepts developed over a century ago. In 2013, the Nobel Prize in Chemistry was awarded jointly to three pioneers of molecular dynamics – Martin Karplus, Michael Levitt, and Arieh Warshel – for the development of multiscale models for complex chemical systems. Recently, the power of MD simulations has been substantially improved in terms of modeling size and time length. With rapid progress in computational hardware and software, researchers are now able to study a broad range of molecular systems and condensed matter under different conditions. MD simulations have found extensive applications in fields such as materials science, biochemistry, and drug discovery to gain insights into the behavior of molecular systems that are difficult or impossible to study experimentally. The simulations can investigate the properties of materials ranging from molecules, polymers, liquids, and solids to various biomolecular systems such as proteins, nucleic acids, and membranes. One prominent achievement in recent MD simulations is the computational modeling of the whole SARS-CoV-2 virion [19]. The simulated model contains a total of 305 million atoms! The whole system was simulated on the Summit supercomputer at the Oak Ridge National Laboratory (ORNL) for a total time of 84 ns, with the aid of various state-of-the-art simulation toolkits, machine learning (ML) methods, and GPU acceleration technology.

To carry out MD simulations, the potential energy function $V(\mathbf{r})$ of a molecular system needs to be determined based on empirical MM methods. With this potential energy function, the force $\mathbf{F}$ that acts upon an object in the molecular system can be obtained as the negative of the first derivative of this function

$$F = -\nabla V(r). \tag{6.16}$$

Once the force F is known, the position changes of the object in time can be further determined using the classical Newton's equations of motion as described in Section 6.2. A typical MD simulation workflow is depicted in Figure 6.22. With the potential energy function and initial positions of a molecular system as inputs, the MD simulations allow the positions and velocities of all the atoms in the simulated system to evolve through successive time steps. Analysis of the outputs of MD simulations provide useful information for understanding various physical and chemical phenomena, such as fluctuations of molecular structures and conformations, thermodynamic properties, diffusion behavior, and reaction kinetics. Moreover, MD simulations can guide experimental work by providing quantitative understanding of how the systems work at the molecular level. This methodology has been popularly applied in the research areas of functional materials design, spectroscopic analyses, and drug discovery, just to name a few.

In practice, the execution of an MD simulation task is not easy for a beginner. A certain level of theoretical background knowledge in combination with various computer skills are needed. Unlike the examples of QM calculations demonstrated above, an MD simulation job requires more detailed specifications of a wide range of parameters, including force fields, boundary conditions, degrees of freedom (constraints), temperature, pressure, and solvent models. Interested readers are recommended to study relevant textbooks as well as to take MD training workshops offered by experts. There are many MD simulation programs available, some of which are licensed, and others are free to general or academic users. Among them, *AMBER*, *Discovery Studio*, *GROMACS*, *LAMMPS*, and *NAMD* are

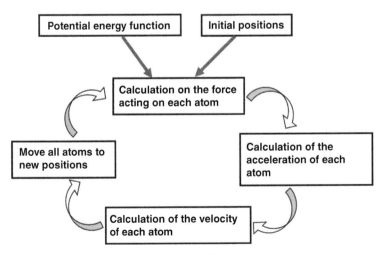

Figure 6.22 A general flow diagram of MD simulations.

quite popular in the MD simulation community. Each of these programs shows excellent performance in modeling certain types of systems but may not be suitable for others. The choice of an MD simulation program is therefore dependent on the research purposes and the capacities of the software.

6.5.4 A Case Study of *Ab Initio* Molecular Dynamics Simulations

Besides classical MD simulations, a methodology based on *ab initio* calculations has recently gained increasing popularity in computational organic chemistry. One of the key differences between the *ab initio* molecular dynamics (AIMD) and the classical MD approaches is that AIMD solves the Schrödinger equation and therefore is based on real physical potentials. As such, AIMD can provide more accurate and unbiased simulations for chemical processes. For example, the electron polarization effects are directly included in AIMD calculations, allowing chemical reactivity (e.g., bond forming and breaking) to be modeled more effectively than the classical MD methods. In the past few decades, AIMD simulations have been rapidly developed as an important theoretical tool for understanding organic reaction mechanisms, excited-state properties, and catalytic effects in the solid and liquid phases.

Most of the AIMD simulations are achieved under the theoretical framework of the Born-Oppenheimer (BO) or Car-Parrinello (CP) schemes [20,21]. In the BO scheme, the equations that describe the wavefunctions are solved at each time step. In the CP scheme, however, the wavefunction parameters are propagated as classical degrees of freedom, so that repeatedly solving the electronic wavefunctions is avoided. The BOMD and CPMD methods are implemented in various QM programs, such as *CPMD*, *CP2K*, *Gaussian*, and *ORCA*. In this case study, a simple AIMD input file for *ORCA* (ver. 5.04) is illustrated in Figure 6.23. The *ORCA* program runs AIMD simulations under the BO scheme.

```
! MD wB97X-D3 6-31+G(d)
%md
   initvel 300_K
   timestep 0.5_fs
   thermostat berendsen 300_K timecon 10.0_fs
   dump position stride 1 filename "trajectory.xyz"
   run 500
end
%pal
 nprocs 32
end
%maxcore 4000
* xyz -1 1
(Initial Cartesian Coordinates)
*
```

Figure 6.23 An example of AIMD simulation input file for *ORCA*.

In this example, a BOMD simulation job is to run, starting from the 10th structure extracted from the previous relaxed PES scan calculations shown in Figure 6.13. In this structure, the cyanide carbon exhibits a distance of 2.07 Å to the carbonyl carbon, which is close to the geometry of the calculated transition state (see Figure 6.14). In the first line of the input file, the keyword "**MD**" denotes an AIMD job to run. The DFT method uses the ωB97X-D3 functional and the 6-31+G(d) basis set. Lines 2 to 7 describe detailed parameters for the AIMD calculations. In line 3, "**initvel 300_K**" sets the initial velocities according to a temperature at 300 K. In line 4, the keyword "**timestep**" denotes the time interval for each step of simulation. Timestep determines the computational expense and accuracy of MD simulations. Short time steps give better accuracy but prolong the computational time. Large time steps speed up the simulation but may cause the simulation to be unstable. In this case study, a time step of 0.5 fs is chosen. In the next line of the input file, a Berendsen-type thermostat is specified for use, by which the simulation temperature is kept at 300 K. In MD simulations, a **thermostat** is used to regulate the average temperature of the simulated system in some fashion. Typically used thermostats are Anderson, Berendsen, Gaussian, Langevin, and Nosé-Hoover thermostats. MD simulations can be done with various thermodynamic ensembles. The *ORCA* program uses a Velocity Verlet algorithm to solve the equations of motion, and by default the simulations are completed using the canonical or constant NVT ensemble, in which the number of particles (N), the volume (V), and the temperature (T) are kept constant. In line 6, the keyword "**dump**" lets the program periodically generate updated structures. In this MD job, every one step of simulation and the results are saved in a file that is named "**trajectory.xyz**." After simulations are completed, this trajectory file contains all the MD simulated Cartesian coordinates for visualization and analysis. In line 7, the keyword "**run**" specifies how many steps of MD simulations to perform. Herein, 500 steps are set for this job, which is equivalent to a total simulation time of 250 fs, given that each time step is 0.5 fs.

After the job is done, the *ORCA* program generates a series of output files containing useful structural and energetic details resulting from the AIMD simulations. Figure 6.24A shows the correlation plots of the simulation time with the total energy (E_{tot}) and the C–C distance (d_{C-C}) between the cyanide anion and the carbonyl carbon, respectively. According to the energy plot, the system arrives at the highest-energy position at 62 fs. The simulated structure at this time shows a d_{C-C} value of 1.99 Å, which is close to that of the optimized transition state. At 150 fs, the simulated structure shows a d_{C-C} value of 1.51 Å, indicating the formation of a typical C–C single bond between the cyanide and carbonyl carbon. Overall, this simple AIMD simulation reasonably delineates a dynamic trajectory for the nucleophilic addition of cyanide anion on (S)-2,3,3-trimethylbutanal via the Felkin-Ahn transition state.

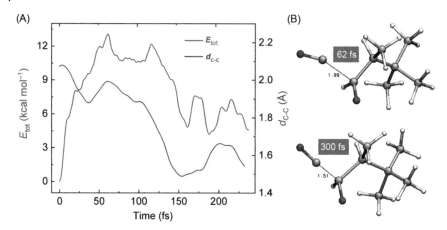

Figure 6.24 (A) Plots of the total energy (E_{tot}) and the distance between the cyanide and carbonyl carbons (d_{C-C}) as a function of AIMD simulation time. (B) Snapshots of molecular structures at different simulation times.

6.5.5 A Case Study of Tight-binding MD Simulations

Empirically derived force fields can be routinely applied in classical MD simulations on very large systems (e.g., proteins, membranes) and a relatively long timescale (nanoseconds to milliseconds). On the other hand, they face limitations such as challenging calculations of nonbonded interactions and the lack of force fields transferable for general purposes. AIMD provides accurate and detailed information about reactivity at the molecular level, but this approach is computationally expensive owing to the need for solving the Schrödinger equation. Therefore, AIMD simulations can only tackle relatively small molecular systems and simulate on a much shorter time scale compared with classical MD simulations. As discussed above, semi-empirical QM methods bridge the gap between the *ab initio* QM and force field methods, since the semi-empirical methods are established upon more drastic approximations than the Hartree-Fock or Kohn-Sham DFT methods and thus run at least two or three orders of magnitude faster than they do. In the meantime, semi-empirical methods are still based on first principles and hence show a wider scope of applications than force field methods. The use of semi-empirical methods comes at the price of significantly reduced accuracy and robustness. Nonetheless, if parameters are carefully adjusted, they can yield sufficiently good results for molecular modeling and simulations.

The development of semi-empirical QM methods has a long history, dating back to the 1970s. The past two decades have witnessed a renaissance of semi-empirical QM methods, mainly as a result of the development of density functional tight binding (DFTB) methods [22,23]. Nowadays, tight binding methods have been widely used as a simple and computationally very fast tool in organic chemistry, material simulations, and biological modeling. In this section, the application of a class of extended tight-binding (xTB) QM methods [24] in simulating non-covalent interactions is demonstrated. The xTB methods have been included in various state-of-the-art quantum chemistry programs such as *CP2K, ORCA,* and *TeraChem.*

In Chapter 3, a molecule termed "buckycatcher" [25–27] and its binding with C_{60} fullerene are discussed (see Figure 3.10). In this case study, the *xTB* program (xtb-docs.readthedocs.io) developed by Grimme and co-workers is used to simulate the binding processes. Note that the *xTB* program is free to general users and runs on the Linux platform. An MD simulation task can be carried out easily by typing the command "**xtb buckycatcher.xyz -input md.inp -omd**" on the Linux terminal. Two input files need to be prepared before the MD simulation. The first file is called **buckycatcher.xyz**, which contains the pre-set Cartesian coordinates of a buckycatcher molecule and a C_{60} molecule. The second file "**md. inp**" includes the conditions for the MD simulations.

Figure 6.25 shows the detailed contents of the two input files. In the .xyz file that contains the Cartesian coordinates of the system, the first line must give the total number of atoms and the second line is for annotation. Starting from the third line, the atomic symbols and their x,y,z coordinates (in the unit of Å) are listed. In this example, the coordinates are set in a way that the C_{60} molecule is positioned 10.45 Å above the bottom of the cavity created by the buckycatcher, leaving no significant non-covalent contact between them (see Figure 6.26). In the .inp file,

buckycatcher.xyz

```
148

C     2.2608064    -1.2498505    -10.1534450
C     1.4378197    -0.5459577     -9.2520013
C     1.4907485     0.8627529     -9.2342419
C     2.1962492     1.6113458    -10.1838327
C     2.9324872     0.8602164    -11.0985183
C     3.0327622    -0.5408688    -11.0522310
C     0.7283471     1.6821866     -8.2407631
C     1.3864485     2.4930319     -7.3288453
C     0.7281491     3.3052191     -6.4154499
C    -0.7446922     3.2454104     -6.3432030
C    -1.4223264     2.4778788     -7.2848252
C    -0.6821907     1.6376738     -8.1666094
C     1.4555931     4.0779849     -5.4353852
C     0.7412512     5.0456805     -4.7503989
C    -0.6914647     4.9983402     -4.7845429
........
```

md.inp

```
$md
   temp=298.15  # in K
   time= 100.0   # in ps
   dump= 100.0   # in fs
   step=   4.0   # in fs
   velo=false
   nvt =true
   hmass=4
   shake=2
   sccacc=2.0
$end
```

Figure 6.25 Two input files for the MD simulations of the binding of buckycatcher and C_{60} fullerene using the *xTB* program.

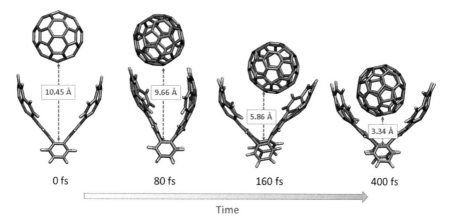

Figure 6.26 Snapshots of a buckycatcher molecule and a C$_{60}$ fullerene taken at varied MD simulation times.

lines 3 to 6 specify the simulation temperature, total time of simulation, interval for trajectory print out, and time step for MD propagation, respectively. In line 7, the keyword "**velo = false**" is the default setting for writing out velocities. In line 8, "**ntv = true**" sets up the simulation in the NTV ensemble. In line 9, "**hmass = 4**" specifies the mass of hydrogen atom to be four times heavier than its typical mass. This specification is used for attaining reasonable simulation stability when a long time step (e.g., 4 fs) is used. Lines 9 and 10 specify the algorithm to constrain bonds and the accuracy of xTB calculations in dynamics.

The *xTB* program generates various output files after the MD calculations, providing detailed trajectories, energies, and other related information. The trajectory file can be visualized using visualization software such as *VMD*. The snapshots in Figure 6.26 clearly show a process where the buckycatcher molecule opens its π-cavity to capture a C$_{60}$ molecule in a period of 400 fs. After 400 fs, the buckycatcher hosts the C$_{60}$ molecule through π–stacking between the two corannulene arms and C$_{60}$.

6.6 Machine Learning in Nanochemistry

Machine learning (ML) has recently emerged as an important computational technique in many fields of science and engineering. ML belongs to a subcategory of the field of **artificial intelligence** (AI) and it is specifically aimed at pattern recognition. ML algorithms are computer programs that learn to carry out specific tasks through inspection of a data bank without explicit human instructions. It is especially powerful in handling a body of data that is too big and complex for human researchers to process using traditional methods.

In today's chemical research, the application of high-throughput experimental and computational methods has generated enormous amounts of as well as a large variety of different types of scientific data, ranging from molecular properties, reaction yields and conditions, to spectroscopic details and QM-computed molecular structural and dynamic properties. With the aid of ML algorithms, it is now possible to analyze these data by training models to classify observations into discrete groups, to learn which features determine a particular metric of performance (e.g., catalytic activity, electronic bandgap), and to predict the outcomes of new experiments. Indeed, ML has significantly reshaped the ways of collecting, analyzing, and interpreting data in many scientific disciplines.

ML uses algorithmic approaches, ranging from basic regression techniques to state-of-the-art methods, to achieve classification of data, identification of empirical correlations within the data, and prediction of the consequences of these correlations for new data. The ML algorithms learn from the data itself to refine the accuracy of their predictions typically by minimizing a mathematical error function. In ML, algorithms are written to address classes of problems and then trained for a specific task based on which type of data is available. ML in chemical research mainly uses the **supervised learning** approach. In supervised learning, the data are labeled, meaning that each piece of data comprises an input (e.g., parameters of an experiment or design of a material) and an output (e.g., outcome of an experiment or material property of interest). The algorithm, for example, an artificial neural network (Figure 6.27), takes in the input features of the data set and builds a model based on internal assumptions to produce the output of the data set with as little error as possible. This model, which is just a mapping of input to output, can then be used to predict the outputs when it is given inputs that are not included in the training data set. For supervised learning, the input features of the data must be pre-determined and are not selected by the algorithm.

Prediction and classification are two tasks commonly performed by supervised ML. If there is no information about output in the data, **unsupervised learning** can be used to uncover relationships. Clustering and component analysis are

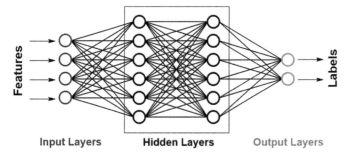

Figure 6.27 Schematic illustration of a simple neural network.

tasks that are commonly performed via unsupervised learning. It is also possible to carry out **semi-supervised** ML, in which the model is initially established by supervised learning on labeled data and then refined by unsupervised learning on unlabeled data. Through the extensive work of mathematicians and computer scientists, algorithms for dealing with common problems can be readily used in an off-the-shelf manner from diverse open-source platforms.

In a chemistry ML project, a list of candidate molecules needs to be collected at first, which are ideally to be represented by their 3D structures (e.g., Cartesian coordinates). However, this method consumes too much storage space if the number of molecules in the data set is large. It is therefore common for a list of molecules to be presented in a somewhat compressed format, known as **chemical representations**. A chemical representation reduces the dimensionality of the 3D structure into a chemically meaningful format, such as a chemical formula, a structural formula, SMILES, SMARTS, and InChI. To chemists, both chemical formulas and structural formulas are very familiar. Chemical formulas, such as H_2O and CH_3CH_2OH, are the simplest way to represent a molecule, but they lack detailed structural information. Compared with chemical formulas, structural formulas provide a much better way of representing molecules. They are therefore frequently used in chemistry textbooks and the literature. Structural formulas retain considerably more structural information than chemical formulas. A trained chemist can easily visualize the 3D structure of a molecule by glancing at its structural formula. In ML, however, structural formulas are not an ideal format for storing molecular structures, as they require the data to be stored in an image format and hence are space consuming if the number of molecules is vast. Instead, some ML-friendly formats of chemical representations (e.g., a string of ASCII characters) are preferred.

SMILES is a type of chemical representation that strikes a balance between chemical formulas and structural formulas [28]. SMILES stands for **S**implified **M**olecular **I**nput **L**ine **E**ntry **S**ystem. It is a chemical notation that uses a short ASCII string to describe molecules. As such, SMILES can be easily generated, read, and translated by a computer. In practice, researchers do not need to worry about generating ambiguous SMILES strings, as the computer software can automatically check them and assign a unique SMILES string to each of the structures when necessary. SMILES uses a number of syntax rules. For example, atoms are written as their atomic symbols. Single, double, triple, and quadruple bonds are written as "–," "=," "#," and "$," respectively. A branch from a chain is placed between a pair of parentheses. Atoms in aromatic rings are written as lowercase letters. Reactants and products are separated by the symbol ">," and so on. For example, the SMILES of 2-propanol is written as "CC(O)C," and acetophenone is "CC(=O)c1ccccc1."

SMARTS is an extension of SMILES. In fact, almost all SMILES specifications are valid SMARTS targets. The use of SMARTS allows researchers to make more flexible and efficient substructure-search specifications. InChI (International

Chemical **I**dentifier) is a structure-based chemical identifier, which was originally developed by IUPAC [29]. As a standard identifier for chemical databases, InChI is essential for effective information management across chemistry. In common practice, InChI is used as a unique identifier and SMILES for storage and interchange of chemical structures.

In a supervised ML setting, a model is constructed in the form of $y = f(\mathbf{x})$, where y is called **response variable** (possibly a vector) and $\mathbf{x}$ denotes **predictor variables**. In chemistry, the response variable can be a physical or chemical property of interest (e.g., intrinsic solubility, HOMO/LUMO energy, or reactivity), while the predictor variables are elements in the structures of molecules. Before building an ML model, a subset of molecules needs to be selected for training and testing. To do so, the response variable of each molecule in this subset must be obtained either through experimental measurements or high-level QM calculations (e.g., DFT). Aside from this, a set of features to describe the molecules must be decided. Such features are structural variables that are related to the physical or chemical properties of interest. However, there is no systematic method for constructing the features of molecules. Strategies developed in the literature are always helpful for consideration and decision. After the above steps, one may start to apply ML models on the basis of supervised learning. There are many families of ML models, which are well known by their acronyms (e.g., ANN, CNN, SVM, and SVR). To determine an appropriate ML model for chemical research, understanding of the algorithms as well as insights into the data are important [30–33]. The selected model is trained on the subset of data, known as the training data set, to learn the relationships between the features and response variables. The trained model can be further evaluated by another set of data, known as the validation data. Finally, the model is subjected to optimization by adjusting relevant hyperparameters. Once an ML model has been evaluated and optimized, it can be deployed to make predictions on new data. In chemistry, this could involve predicting the properties of new functional materials, unprecedented drug molecules, or new synthetic routes and reaction conditions. It is also possible to use a well-trained ML model to guide autonomous research. For example, a term **self-driving laboratory** (SDL) has been proposed, which describes an AI-powered lab that carries out research activities from generation of hypotheses to experimental tests without human intervention.

In a summary, the application of ML in chemistry has already changed the ways in which chemists conduct experimental research and data analysis. In the future, ML and other AI technologies are expected to continuously influence and reshape many areas of chemistry. In view of the multidisciplinary nature of nanochemistry, it is compelling for researchers in the field to embrace the rapid advancements of ML and other AI techniques. It is anticipated that ML will become a powerful and indispensable tool in future chemical research, for example, accelerated discovery of new nanomaterials, improved drug search, enhanced understanding of complex reaction mechanisms, and highly efficient self-driving chemistry labs.

Further Reading

- Bachrach, S. M. Computational Organic Chemistry. 2nd ed.; John Wiley & Son: Hoboken, New Jersey, 2014.
- Ratner, M. A.; Schatz, G. C. Introduction to Quantum Mechanics in Chemistry. Prentice Hall: Upper Saddle River, New Jersey, 2001.
- Simons, J.; Nichols, J. Quantum Mechanics in Chemistry. Oxford University Press: New York, 1997.
- Seminario, J. M.; Politzer, P. Modern Density Functional Theory: A Tool for Chemistry. Elsevier: Amsterdam, 1995.
- Rapport, D. C. The Art of Molecular Dynamics Simulation. Cambridge University Press: Cambridge, 1995.
- Cartwright, H. M. Machine Learning in Chemistry: The Impact of Artificial Intelligence: The Impact of Artificial Intelligence. Royal Society of Chemistry: Cambridge, 2020.

References

1 Cram, D. J.; Elhafez, F. A. A., Studies in Stereochemistry. X. The Rule of "Steric Control of Asymmetric Induction" in the Syntheses of Acyclic Systems. *J. Am. Chem. Soc.* **1952**, *74*, 5828–5835.

2 Chérest, M.; Felkin, H.; Prudent, N., Torsional Strain Involving Partial Bonds. The Stereochemistry of the Lithium Aluminium Hydride Reduction of Some Simple Open-Chain Ketones. *Tetrahedron Lett.* **1968**, *9*, 2199–2204.

3 Bürgi, H.; Dunitz, J.; Shefter, E., Geometrical Reaction Coordinates. II. Nucleophilic Addition to a Carbonyl Group. *J. Am. Chem. Soc.* **1973**, *95*, 5065–5067.

4 Perdew, J. P.; Schmidt, K. In *Jacob's Ladder of Density Functional Approximations for the Exchange-Correlation Energy.* AIP Conf. Procd., Am. Inst. Phys. **2001**, *577*, 1–20.

5 Morgante, P.; Peverati, R., The Devil in the Details: A Tutorial Review on Some Undervalued Aspects of Density Functional Theory Calculations. *Int. J. Quant. Chem.* **2020**, *120*, e26332.

6 Cohen, A. J.; Mori-Sánchez, P.; Yang, W., Challenges for Density Functional Theory. *Chem. Rev.* **2012**, *112*, 289–320.

7 Verma, P.; Truhlar, D. G., Status and Challenges of Density Functional Theory. *Trends Chem.* **2020**, *2*, 302–318.

8 Neese, F.; Wennmohs, F.; Becker, U.; Riplinger, C., The ORCA Quantum Chemistry Program Package. *J. Chem. Phys.* **2020**, *152*, 224108.

9 Hoffmann, R., Extended Hückel Theory—V: Cumulenes, Polyenes, Polyacetylenes and Cn. *Tetrahedron* **1966**, *22*, 521–538.

10 Kaiser, K.; Scriven, L. M.; Schulz, F.; Gawel, P.; Gross, L.; Anderson, H. L., An sp-Hybridized Molecular Carbon Allotrope, Cyclo[18] carbon. *Science* **2019**, *365*, 1299–1301.

11 Baryshnikov, G. V.; Valiev, R. R.; Kuklin, A. V.; Sundholm, D.; Ågren, H., Cyclo[18] Carbon: Insight into Electronic Structure, Aromaticity, and Surface Coupling. *J. Phys. Chem. Lett.* **2019**, *10*, 6701–6705.

12 Liu, Z.; Lu, T.; Chen, Q., An sp-Hybridized All-Carboatomic Ring, Cyclo[18] carbon: Bonding Character, Electron Delocalization, and Aromaticity. *Carbon* **2020**, *165*, 468–475.

13 Schleyer, P. v. R.; Maerker, C.; Dransfeld, A.; Jiao, H.; van Eikema Hommes, N. J., Nucleus-Independent Chemical Shifts: A Simple and Efficient Aromaticity Probe. *J. Am. Chem. Soc.* **1996**, *118*, 6317–6318.

14 Fallah-Bagher-Shaidaei, H.; Wannere, C. S.; Corminboeuf, C.; Puchta, R.; Schleyer, P. v. R., Which NICS Aromaticity Index for Planar π Rings Is Best? *Org. Lett.* **2006**, *8*, 863–866.

15 Chen, Z.; Wannere, C. S.; Corminboeuf, C.; Puchta, R.; Schleyer, P. v. R., Nucleus-Independent Chemical Shifts (NICS) as An Aromaticity Criterion. *Chem. Rev.* **2005**, *105*, 3842–3888.

16 Fernandez, I. *Aromaticity: Modern Computational Methods and Applications.* Elsevier: Armsterdam, **2021**.

17 Marques, M. A.; Ullrich, C. A.; Nogueira, F.; Rubio, A.; Burke, K.; Gross, E. K. *Time-Dependent Density Functional Theory.* Springer-Verlag: Berlin, **2006**.

18 Shi, B.; Yuan, L.; Tang, T.; Yuan, Y.; Tang, Y., Study on Electronic Structure and Excitation Characteristics of Cyclo[18]carbon. *Chem. Phys. Lett.* **2020**, *741*, 136975.

19 Casalino, L.; Dommer, A. C.; Gaieb, Z.; Barros, E. P.; Sztain, T.; Ahn, S.-H.; Trifan, A.; Brace, A.; Bogetti, A. T.; Clyde, A., AI-driven Multiscale Simulations Illuminate Mechanisms of SARS-CoV-2 Spike Dynamics. *Int. J. High Perform. Comput. Appl.* **2021**, *35*, 432–451.

20 Hutter, J., Car–Parrinello Molecular Dynamics. *Wiley Interdiscip. Rev. Comput. Mol. Sci.* **2012**, *2*, 604–612.

21 Marx, D.; Hutter, J. *Ab Initio Molecular Dynamics: Basic Theory and Advanced Methods.* Cambridge University Press: Cambridge, **2009**.

22 Seifert, G.; Joswig, J. O., Density-Functional Tight Binding—An Approximate Density-Functional Theory Method. *Wiley Interdiscip. Rev. Comput. Mol. Sci.* **2012**, *2*, 456–465.

23 Koskinen, P.; Mäkinen, V., Density-Functional Tight-Binding for Beginners. *Comput. Mater. Sci.* **2009**, *47*, 237–253.

24 Bannwarth, C.; Ehlert, S.; Grimme, S., GFN2-xTB—An Accurate and Broadly Parametrized Self-Consistent Tight-Binding Quantum Chemical Method with Multipole Electrostatics and Density-Dependent Dispersion Contributions. *J. Chem. Theor. Comput.* **2019**, *15*, 1652–1671.

25 Sygula, A.; Fronczek, F. R.; Sygula, R.; Rabideau, P. W.; Olmstead, M. M., A Double Concave Hydrocarbon Buckycatcher. *J. Am. Chem. Soc.* **2007**, *129*, 3842–3843.

26 Mück-Lichtenfeld, C.; Grimme, S.; Kobryn, L.; Sygula, A., Inclusion Complexes of Buckycatcher with C_{60} and C_{70}. *Phys. Chem. Chem. Phys.* **2010**, *12*, 7091–7097.

27 Le, V. H.; Yanney, M.; McGuire, M.; Sygula, A.; Lewis, E. A., Thermodynamics of Host–Guest Interactions between Fullerenes and a Buckycatcher. *J. Phys. Chem. B* **2014**, *118*, 11956–11964.

28 Weininger, D., SMILES, a Chemical Language and Information System. 1. Introduction to Methodology and Encoding Rules. *J. Chem. Infor. Comput. Sci.* **1988**, *28*, 31–36.

29 Heller, S. R.; McNaught, A.; Pletnev, I.; Stein, S.; Tchekhovskoi, D., InChI, the IUPAC International Chemical Identifier. *J. Cheminform.* **2015**, *7*, 1–34.

30 Artrith, N.; Butler, K. T.; Coudert, F.-X.; Han, S.; Isayev, O.; Jain, A.; Walsh, A., Best Practices in Machine Learning for Chemistry. *Nature Chem.* **2021**, *13*, 505–508.

31 Butler, K. T.; Davies, D. W.; Cartwright, H.; Isayev, O.; Walsh, A., Machine Learning for Molecular and Materials Science. *Nature* **2018**, *559*, 547–555.

32 Meuwly, M., Machine Learning for Chemical Reactions. *Chem. Rev.* **2021**, *121*, 10218–10239.

33 Bender, A.; Schneider, N.; Segler, M.; Patrick Walters, W.; Engkvist, O.; Rodrigues, T., Evaluation Guidelines for Machine Learning Tools in the Chemical Sciences. *Nature Rev. Chem.* **2022**, *6*, 428–442.

Index

Organic Nanochemistry: From Fundamental Concepts to Experimental Practice,
First Edition. Yuming Zhao.
© 2024 John Wiley & Sons, Inc. Published 2024 by John Wiley & Sons, Inc.

Printed and bound by CPI Group (UK) Ltd, Croydon, CR0 4YY

16/04/2025

14658415-0001